MOJAVE DESERT VOICES:

THE JARAMILLO, OLGUIN, YSLAS, LUNA AND ARREDONDO FAMILIES

MOJAVE DESERT VOICES:

THE JARAMILLO, OLGUIN, YSLAS, LUNA AND ARREDONDO FAMILIES

To the Perez Family — Clifford J Walker

Clifford J. Walker, Editor

A MOJAVE RIVER VALLEY MUSEUM PUBLICATION
BARSTOW, CA 92312-1282

Mojave River Valley Museum Association
270 Virginia Way
P.O. Box 1282
Barstow, Ca 92311-1282

Edited by Clifford J. Walker

Set in 11 point Garamond type

Design and Layout by Ron McKinley
Southwest Savvy, Apple Valley
Cover Design by Maggie Heyn, Norco, CA

Five Families Committee: Pita Jaramillo, Naydean Jaramillo Williams, Gillie Olguin, Elaine Olguin Villareal, Katy Yslas-Yent, Vernon "Flip" Williams, Leo Luna, Helen Luna Raster, Patsy Luna Biles, Ruben Arredondo, Cliff Walker, Ted and Marsha Weasma, LaVella Tomlinson and Martha Burnau.

Library of Congress Catalog Card Number 2001096313

Mojave Desert Voices: The Jaramillo, Olguin, Yslas, Luna and Arredondo Families

A Mojave River Valley Publication
Includes Index

Mojave Desert History 1917-2001. 2. Barstow, CA. 3. Mexican-American Families in the Desert. 4. Life in a Railroad Town. 5. New Mexican Connection. 6. Route 66 Stories

ISBN 0-918714-09-5

PREFACE

Full many a flower is born to blush unseen
And waste its sweetness on the desert air.
Full many a gem of purest ray serene
The dark unfathomed caves of ocean bear.

—Excerpt from *Elegy Written in a Country Churchyard* by Thomas Gray

The Mojave River Valley Museum would like you to view some beautiful flowers that have been growing in the Mojave Desert, to smell their sweetness. We want to introduce some remarkable gems on the Mojave Desert, people who have gone through some hard times, and tragedies, but worked hard, sometimes two-three jobs, and raised their children with values of industry, honesty, and especially love of family, God, community and country.

And the marvelous thing is this: These desert communities are filled with hundreds of remarkable families—all with unique stories, all a part of their communities' histories. Indeed the museum needs their stories to make the history of the desert more complete.

In the past, the museum has honored several families during the Annual Museum BBQs, under the leadership of local historians Germaine Moon and Patricia Keeling. They honored the Mulcahys, a mining family, the Hills of Community Boulevard, a ranching and dairy family, and the Ramirez family, desert ranchers since the 1910s. From the research and photographs of these families, the museum has added much to the history of the Mojave Desert, capturing vital heritage that may have been lost to time. Heritage thus preserved for the future.

The museum had already archived the Arredondo family photographs, and before Ramon passed away two years ago, he and brother Ruben (Bennie) wrote part of their family's history. Because the Arredondos have been in the Mojave Desert since 1915, we wanted to write a booklet including their family. Since the 35th Annual Museum BBQ in 2001, came on May 5, Cinco de Mayo, the museum decided to honor five Latino families at the BBQ. The Arredondos would be one. Four other families, Jaramillo, Olguin, Yslas and Luna, had begun to collect photographs and to tell stories of their Route 66 connections, between the High Desert, Arizona and New Mexico. They were eager to write about their past and their heritage. The Museum Board voted to honor these four families too, and voted to put their stories into a book of about 120 pages. In order to do their families justice, however, the book needed to be enlarged.

The Five Families Committee would eventually like to build a new wing on the museum to expand it for cultural diversity. The Committee realizes that the five families are special, but each family in the desert is special. They know the museum could have chosen numerous others—all with unique histories, all helping to make the desert a better place to live, all making the United States a stronger country. Since the museum wanted to write the stories of desert families, this seemed like a good place to start. We hope we can continue over the years and write more books of desert family histories.

The museum and the Committee encourage all families to write their histories, to collect and label their family pictures and documents. Interview the older generations, before their gems are lost. These stories are important parts of a community's history—often not written and therefore eventually lost. For assistance, families can receive packets at the museum to get started. Contact any member of the Five Family Committee for advice and suggestions. The Church of Jesus Christ of the Latter Day Saints on Barstow Road will be glad to help families get started. They are gracious and helpful, and are usually open from 9:00 a.m. to 9:00 p.m., Tuesday through Thursday. Finally, use this book as a model of several ways to tell the histories. There is no "right" way. The "wrong" way is to procrastinate and NOT to do it. THEREFORE, WRITE YOUR FAMILY'S STORY.

We hope by reading this book and meeting wonderful Americans who have added so much to their communities, the museum will have a more accurate history of the High Desert. As important, perhaps more important, readers will discover the sweetness of these desert flowers and be aware of these "gems of purest ray serene" and thus realize the High Desert communities are filled with wonderful gems that have made these communities stronger. —Five Families Committee

Cover: front cover photo of Joshua tree and sunset courtesy of Kenton Elliot of Barstow. Front photo of Noberto Luna and John Yslas with daughter Katy. Back cover has the Arredondo family in 1936 and two Luna grandchildren Daniel and Shannon showing their *garajes* (garages).

DEDICATED TO THOSE WHO
MADE US WHAT WE ARE
AND TO THOSE
GENERATIONS THAT FOLLOW

ACKNOWLEDGMENTS

Committee members of the Five Families appreciate those who have worked to save the heritage of the desert for future generations. People who supported the Mojave River Valley Museum over the years enabled the museum to help preserve these wonderful stories and this heritage for the future. Your memberships, BBQ tickets, raffle tickets and donations have made this book possible.

For the book production, we thank Ron McKinley of Southwest Savvy, for the layout and design, Maggie Heyn for the cover, and Mike Hamblin and Steve Smith for the scanning and photographic work.

We appreciate those ancestors who made us what we are. Since most of our ancestors are lost in the obscurity of the past, we know only a few forbearers. But by learning about our parents and grandparents and seeing characteristics and values they have passed on to us, we understand that the unknown ancestors have passed on much to those relatives who raised us.

We are blessed having accepting friends, neighbors, schoolmates, fellow workers, and teachers who treated us equally and gave us a great education–to all of you, thank you much.

We thank the younger generations who contributed to this book, giving their perspectives and insights, making this a more valuable history.

Special thanks to Lorenza Osborn for proofing the Spanish parts.

The Five Families Committee:

Pita Jaramillo, Naydean Jaramillo Williams, Gillie Olguin, Elaine Olguin Villareal, Katy Yslas-Yent, Eugene "Flip" Williams, Leo Luna, Helen Luna Raster, Patsy Luna Biles, and Ruben Arredondo, Cliff Walker, Ted and Marsha Weasma, LaVella Tomlinson, and Martha Burnau.

INTRODUCTION

On Friday and Sunday evenings, thousands of cars whiz through the Mojave Desert, heading to the river, to Las Vegas, Death Valley or Los Angeles—many destinations. Most go right through, not noticing the clear air, or the smells of creosote when it rains, or the pinks in the Desert Mountains, or the shadows as they constantly change the landscape. The tourists certainly do not notice the wonderful families that have evolved here in the desert. Those who have come from this desert experience have spread all over the country. Most still consider the desert their home.

Mojave Desert Voices gives the history of five Latino families, just five of the countless wonderful families from desert communities. *Mojave Desert Voices* is a book on family values and keeping family traditions while living the American dream, a book of love for the family, their religion, their country, and a book showing how love and hard work and steady perseverance overcame economic difficulties, prejudice, and family tragedies.

These families, of course, go back thousands of years, but like most families, they can be traced back only a few generations. The Hispanics from New Mexico, especially those whose roots go back to the Spanish colonization, have tremendous pride of family and country. Their New Mexican roots attach deep to the land and to the families. Juan Lopez Olguin (actually started as Holguin and changed after two generations to Olguin) came from Spain and helped settle New Mexico in 1598, nine years before the founding of Jamestown, Virginia. Thus starting 14 generations in what is now part of the United States (See Appendixes). From New Mexico, Gillie and Vina Olguin moved to Barstow right after World War II. The first Arredondos came to Florida while it was still under Spanish control. Jose Arredondo came from Mexico to work on railroads in 1898 in Texas, Colorado and Arizona, and went back to Mexico, ranching until the Mexican Revolution. In early 1915 Jose came back with his wife and worked for over 45 years for Santa Fe RR at various Section Houses in the Mojave Desert. The Jaramillo Family and the Luna family came from New Mexico and some branches lived there for a couple centuries. The Yslas's came from Arizona. These families have researched their pasts and have captured not only their history but also some history of New Mexico, Mexico and the Mojave Desert.

This book is filled with precious moments, moments that are both unique and universal. Family events, such as falling in love, giving birth, nursing the sick and the dying, and saying goodbye as the loved one leaves this physical world. These experiences—wonderful and sad—are universal experiences of all mankind. Yet these precious moments are so personal, so profound, that they are special to all.

Precious Moments:

...bashful young couples discovering each other, and starting what will become a life-long partnership,

...the father who day after day greeted his newly married daughter with a big hug and tears rolling down his cheeks,

...the son the entire family takes pride in for not letting cerebral palsy stop him from doing whatever he wanted to do,

...the grandmother who was the confidante to her teen grandson,

...the girl who grew up to be mayor of the town that didn't want to sell her father a house,

...the family that wouldn't let their mother be put into a care facility, simply saying, "She took care of us,"

...the mother and grandmother who greeted everyone at the door with, "Come on into the kitchen and have some food" and the father who whispers, "Who's that?"

...two grandparents preparing dinner and dancing together in the kitchen to Mexican music,

...the children as they are awakened in the morning of their birthday with the song *Las Mañanitas*, "This is your day, the blessed day that you were born,"

...the five o'clock Sunday dinners at the grandparents, a tradition for at least three generations,

...the young man who didn't care for school but was a natural mechanic and got his dream at 18 with training and backing to open his own shop, when death cut his life just at the budding stage.

Prejudice

Some moments were not always precious. In the desert, as well as most towns in southern California, there were degrees of prejudice. Often the prejudice was subtle, such as when a Barstow High School business teacher told a bright Mexican student to take practical classes because it wouldn't be worthwhile going to college.

Others prejudices in the desert were blatant: Having to sit in the Mexican side of the Forum Theatre; not being part of the decision making of the community; seeing Latino names in the paper only for arrests; being told the Mexican family qualified for a post WW II home in the Foglesong Addition, but were told "We can't sell you a house there"; Railroad Station families (minorities) being able to go swimming in the Union Pacific swimming pool in Yermo only the day before the pool was to be cleaned; being placed in the Mexican School in the first through third grades during the late 1930s to mid-1940s; having a promotion test score in the 90s and a not being promoted, but an "Anglo" was with a score in the 80s; and being the top of the graduating class and not offered a scholarship. Name-calling, hearing ethnic jokes, judging all Latinos by the actions of a few, hiring discriminations in a few places—all were part of growing up in the desert—as well as in numerous places in southern California. Even some Mexicans resorted to name-calling as they perceived a few of these families as being "too white," being called a *"gavacho,"* a gringo, or told "you think you're too good for us."

Tortilla Diplomacy

Part of the integration that occurred over these years was due to spontaneous "Tortilla Diplomacy." People became friends over freshly-made tortillas. These fragrant and tasty delights brought Latinos, Anglos and Native Americans together at the Amboy Mexican celebration of Mexico's Indepencence, September 16th; they helped integrate the government housing at the

Marine Corps Supply Center in the 1950s; and integrated the neighborhood children in many Barstow streets. A warm tortilla did wonders to welcome a potential in-law to the family. Many life-long friendships and even marriages started with a warm buttered tortilla. Or maybe it was two good people starting to like each—then the relationship was enhanced by a couple of tortillas hot off the stove. What counts is people learned how wonderful other groups were and in a few years the desert communities became integrated and more accepting of all groups.

But through hard work, much studying in school by children, being good neighbors, living their Christian religion, and serving a few warm tortillas—all had an effect on creating a community of brotherhood and accepting and judging folks for how they live, not for the color of the skin, the last name or the religion or the ethnic background.

The younger generations of these families did well in school, became athletes, student body officers, and later as adults became labor stewards, church leaders, political campaign chairs, ministers, businessmen and leaders, boards of trustee members, teachers, and professionals of all types—even mayor of Barstow—Katy Yslas-Yent, now Administrative Assistant to Supervisor Bill Postmus. The patience, industry, high values and abilities paid off, carrying this American Dream into the next two generations.

Read *Mojave Desert Voices* and see the manifestation of these dreams. And see too, a reflection of your own American dreams.

Table of Contents

PREFACE iv

ACKNOWLEDGMENTS viii
INTRODUCTION ix
Table of Contents xiii
Map of Southwest U. S. A. and Northern Mexico xiv
Map of Historic U. S. Route 66 xv
Map of Mexico Showing the Origins of the Five Familes xvi
Part 1 The Jaramillo Family 1
Part 2 The Olguin Family 31
Part 3 The Yslas Family...............................60
Part 4 The Luna Family...............................85
Part 5 The Arredondo Family 130
EPILOGUE 176
APPENDIX A JARAMILLO FAMILY 179
APPENDIX B THE OLGUIN FAMILY .. 184
APPENDIX C THE YSLAS FAMILY 188
APPENDIX D THE LUNA FAMILY 191
APPENDIX E THE ARREDONDO FAMILY 195
GLOSSARY... 198
INDEX 200

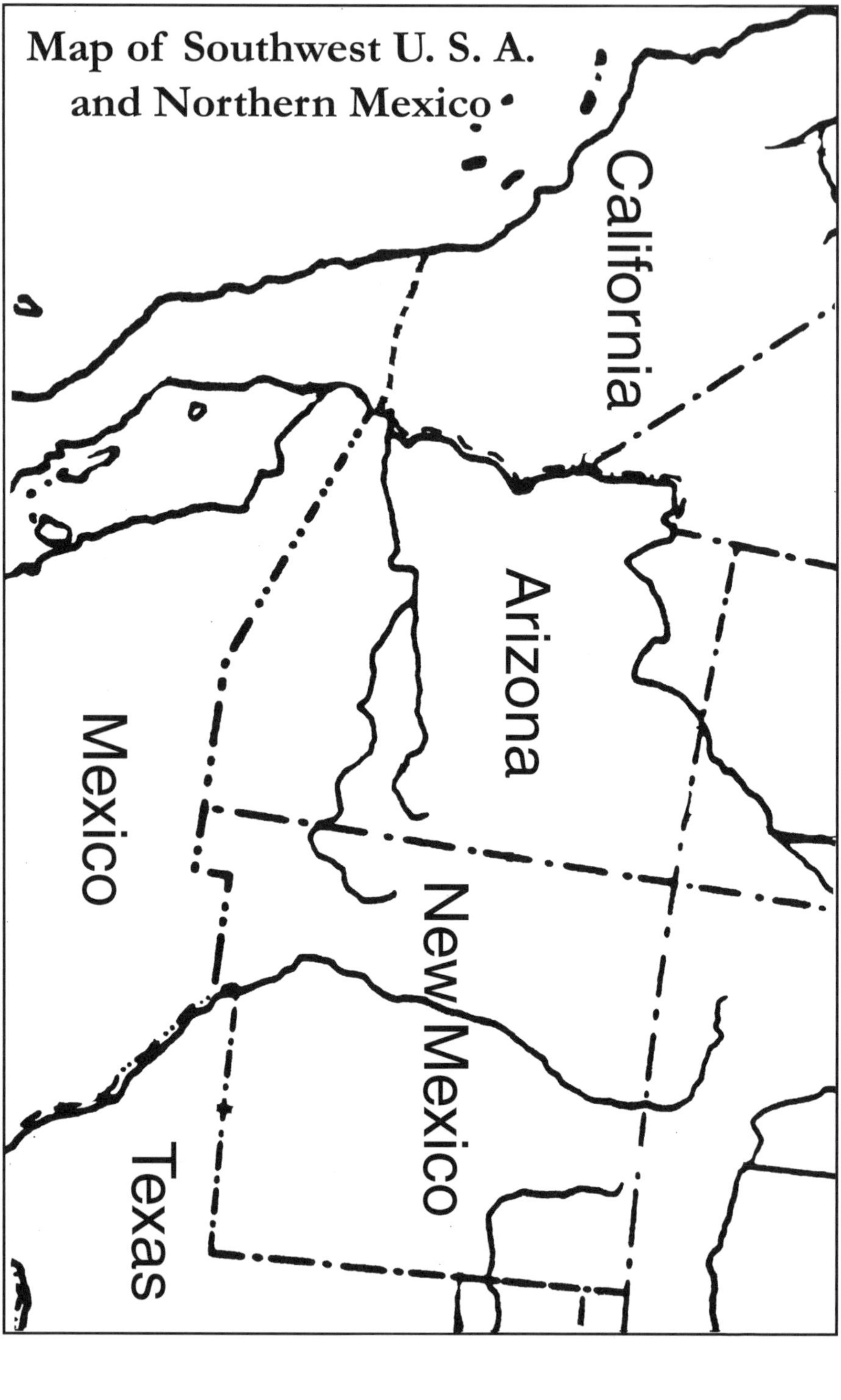
Map of Southwest U. S. A.
and Northern Mexico
California
Arizona
Mexico
New Mexico
Texas

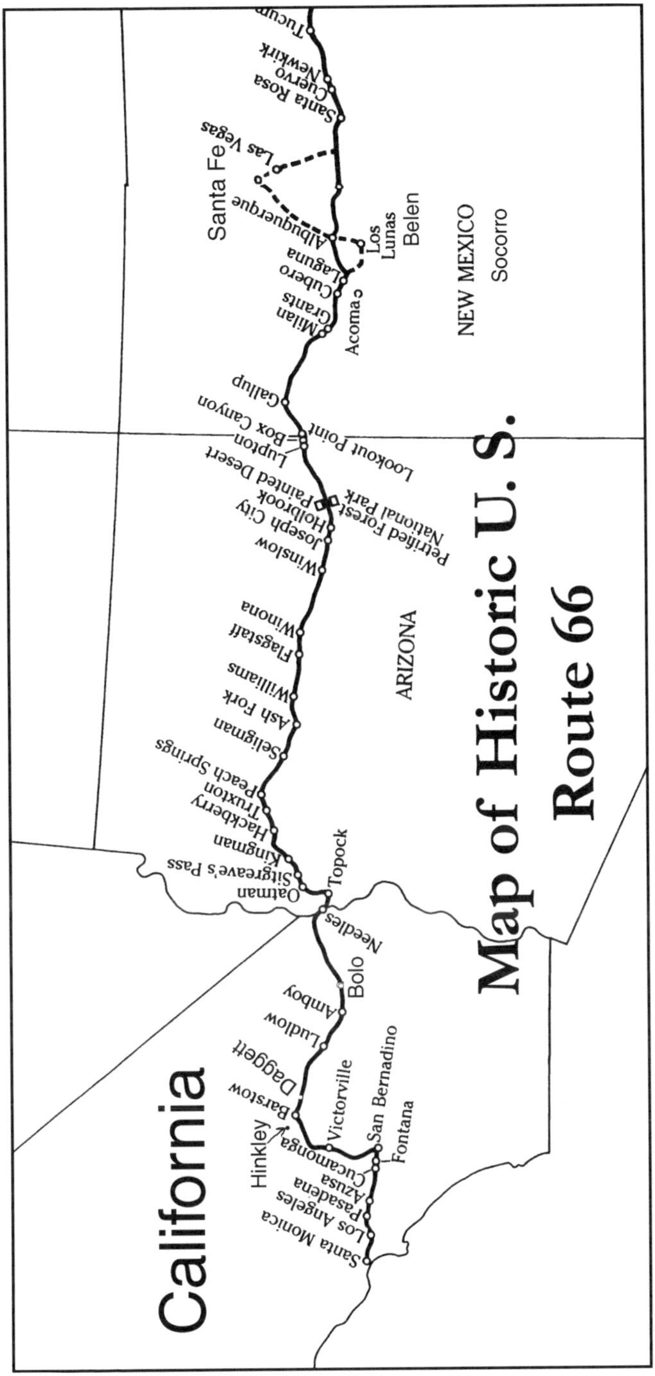

Map of Historic U.S.
Route 66
California
ARIZONA
NEW MEXICO
Santa Monica
Los Angeles
Pasadena
Azusa
Cucamonga
Fontana
San Bernadino
Victorville
Hinkley
Barstow
Daggett
Ludlow
Amboy
Bolo
Needles
Topock
Oatman
Sitgreave's Pass
Kingman
Hackberry
Truxton
Peach Springs
Seligman
Ash Fork
Williams
Flagstaff
Winona
Winslow
Joseph City
Holbrook
Painted Desert
Petrified Forest National Park
Lupton
Box Canyon
Lookout Point
Gallup
Milan
Grants
Acoma
Cubero
Laguna
Albuquerque
Los Lunas
Belen
Socorro
Santa Fe
Las Vegas
Santa Rosa
Cuervo
Newkirk
Tucum

Map of Mexico Showing the Origins of the Five Families

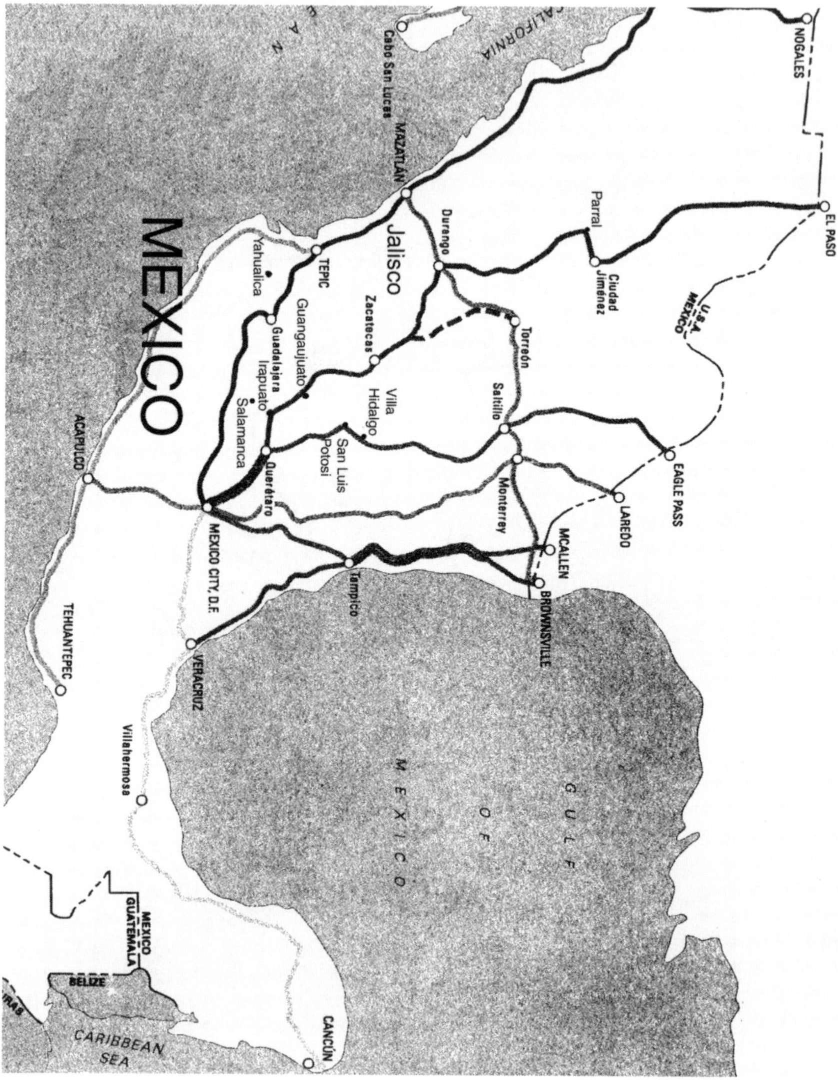

Part 1
The Jaramillo Family
of the
Mojave Desert

Andres N. Jaramillo and Pita as their lives became entwined in New Mexico.

The Jaramillo Story

Dedication to My Husband Andres N. Jaramillo by Pita Jaramillo

My hopes and dreams were shattered when my husband passed away. It took a toll out of me. I could not see myself without him. People still tell me, they can't picture me without my husband. They say that time heals all wounds. It has not worked for me. I was used to seeing him everyday of my life. We were like two peas in a pod. We could not be separated. It is hard to lose a loved one. We were together for over fifty years. You should never take things for granted. As the days go by, you imagine him coming in the door. When you go to bed at night, you feel him right next to you. You can hear him breathing. You extend your hand to touch him, and nobody's there. What a disappointment. What could I have done differently? Maybe he would have lived longer. All these "why's" and no answers. If he only knew how I miss him.

I will always have this empty feeling inside of me. This feeling gets worse from day to day. Being alone does not create loneliness. It's being alone without your loved one that creates loneliness. My husband was a loving person. That's why I miss him so much. I hold on to every memory stored within my being. Our love for each other was overwhelming:

Andres N. and Pita Jaramillo on their wedding day in 1939.

You're the light of my life. I was proud to be

your wife:

Every moment we spent together
Is cherished forever in my heart.

Andres N. Jaramillo was born in Encino, New Mexico. His parents were Nicanor Jaramillo and Maria Efren Bachicha. He had a brother Ted Jaramillo, and sisters Josepheta and Pana (still living at age of ninety in the year of 2000). He was the youngest of the family. His brother and sisters were all married and had children of their own. Andres practically grew up by himself.

Encino was famous for its pinto beans and sheep ranches. This was the distribution center for the marketing of pinto beans. Pinto beans when cultivating do not require any water. Encino water was salty; water for drinking and cooking had to be hauled in. His uncles, the Bachichas, owned most of the ranches. His uncles raised him. The sheep had to be sheered every so often. Since Andy, that's what we called him at home, tagged along, he learned how to sheer the sheep and sack the wool for market. Mutton was their main source of meat. They made jerky out of mutton. The pelts were softened with salt, cleaned and sold.

People from New Mexico were unique. Each town had its own culture and customs. If you happen to come for a visit, no matter what it was, the table would be set for you, with coffee and cookies. The children would get cocoa and cookies. The people from Encino were family oriented. Siblings were taught to be respectful to their elders. When Andy took me to visit his grandmother and aunts and we were ready to leave, he knelt at the feet of his grandmother for a blessing. If perhaps, the elders asked for a drink of water, you would fold your arms, and wait, until the glass was handed back to you.

I was born in Socorro, New Mexico. My life was a bed of roses in comparison to my husband's. We had spring water—the best water in the world. We had two ranches, one in Socorro, the other a few miles from away. We had tenants to take care of the farms. We had chickens, pigs, and cows. In the big ranch, we had all kinds of fruit trees, vegetables, and sugar cane. The most delicious syrup was made out of the sugar cane.

We had so many beautiful trees and flowers in comparison

to the town of Encino which was nothing but plains as far as the eye could see. Winters in Encino it started snowing in September, a very cold country. In Socorro, we had snow in the winter, but not for long. The only hazards we had were the mosquitoes. We have the famous School of Mines. Our Catholic church still stands, exactly as when I was a child. Mt. Caramel School, the school I attended for eight years, still exists. I also boarded at Mt. Caramel.

Andy N. Jaramillo on graduation day in 1937 from high school and father Nicanor Jaramillo and mother Marillita (Maria) Efren Bachicha Jaramillo.

I remember as a child my mother baking. She baked every Friday. We ate the fresh baked bread and sweet breads with homemade butter and jelly. We also indulged on sugar cane syrup with white cheese. My mom would make this white cheese from the cow's milk. During the depression Mom fed all the tramps that came to the door. A lot of people were going hungry. I was still a child, but I remember people coming to the door asking for food. Mom would not turn anyone away.

Pita at her graduation in Albuquerque, New Mexico. The day before Andy asked Pita to marry him, and since she was so young, she answered, "I have to ask my mom."

She said, "We were the fortunate ones. My dad had good job. I remember this old man who rode a donkey; he would come every morning for coffee and sweet bread."

We lived in a two-story house with no neighbors around us. The closest to our house was the Santa Fe Depot with trains going by at a hundred miles an hour. We had lot of tramps getting off of the train. We had two police dogs, to protect us, from vandalism. At that time people were not bad. They would offer to do jobs, around the house, to get something to eat. My mom would feed them, just because they were hungry. Some of them would rather work for their meal. Mom would have them chop wood, which made them feel better. They had earned their meal.

My mom's hobby was taking care of the pigs, chickens and a cow named Bessy. My dad wasn't too much of a ranch hand. He was born in a ranch, lived in a ranch, but did not like the work involved in a ranch. I think he was tired of it. My mom loved it. I had my own chicken that I raised. She had a lot of little chicks. Bessy my cow was such a pet. She would follow me to grazing field, where she would spend the day. When I went in the evening to bring her home she was waiting for me. She would follow me home. I would be eating an apple, and she would slurp down my apple with her big long tongue.

I also had a few kittens, the grandmother, mother and the grand daughter of the cat family. I remember it was Saturday morning, my dad came in and told Mom that the train had run over the daughter kitten. My mom was making my bed, when she shook the blanket, and the little kitten fell to the floor. She had put her little kitten in my bed for safekeeping. My sister was quite a tomboy. She climbed the pole to see the bumblebees that were inside of the pole. She told me to climb the pole and I was bitten all over. My sister didn't even get one bite.

The conductors on the train also spoiled me. They would stop to bring me coal and ice for our home. They would give me a ride in the caboose. I stayed in the train until all the switching of the trains was done. The trackwalker also gave me a ride in his trackwalker to the depot and back.

I also remember churning butter. I would love to see the butter separating from the buttermilk and turn into a ball. This was the good life, no worries *whatsoever.*

My dad traveled quite a bit. I had to board while he was away. My mom went along too. She helped my dad make reports and run the commissary. After graduating from the eighth grade, my dad was transferred to Albuquerque, New Mexico. I started school at St. Mary's High as a freshman.

Andres N. Jaramillo graduated from high school in 1937, and then moved to Albuquerque, New Mexico. Since his mother had just passed away, he lived with his brother, Ted Jaramillo. His bother got him a job in Albuquerque. He also attended school.

I met Andy when I was a sophomore. I worked at Woolworth on weekends and holidays. He would come into the store and buy toothpaste. He came every week and bought the same thing. I asked myself how he could use so much toothpaste. He must have found out that I was going to St. Mary's. I saw him waiting for me. I said to myself, "There is something going on here." I asked him, "How did you know that I was going to school here?"

He said, "I have my ways." Then, he asked me out to dinner.

From that time on we dated. He came to my graduation. The day before graduation he asked me to marry him. I was so young, I told him, "I have to ask my mom." My mother was unhappy about it. She had higher hopes for me.

That very week he came with his brother and sister-in-law to ask for my hand in marriage. My mom told him, "How could you want to marry her? She can't boil water without scorching it."

He said, "I know how to cook."

My dad wasn't too happy either. I was Dad's pet. But Andres was persistent.

We got married at Old Town's San Felipe Catholic Church in Albuquerque, New Mexico, a nice wedding. We went on our honeymoon to Carlsbad Caverns, the most beautiful caverns in the world. These caverns are God's creation.

My mom had us for dinner every night for one whole year. My dad was outside waiting for us. He would hold me in his arms and the tears rolled down his face. I knew I had broken his heart. He wanted me to be with him forever.

We had been married 13 months when our first child was born. It was quite an experience. I didn't have a clue about babies and neither did my husband. Sonny Boy, Andrew Jude, was born. Then, our nightmares started—so many sleepless nights. We thought it was all fun and games. We sure found out the hard way. We had real nice neighbors. She told me to warm some olive oil, and rub olive oil on his tummy. I was willing to do anything to keep him from crying. When I did it, I burned my baby's foot. His foot blistered like balloon. I was crying with my baby. She said, "Child, we better take him to the doctor."

I swore to *myself,* I would never do anything like that again. Experience is your best teacher. My mom always came to the rescue. She finally decided they should move away, so I would grow up and take responsibility. My dad transferred to Hatch, New Mexico.

Eventually I learned to do what had to be done.

My husband helped me quite a bit. He was a better cook than I was. I made biscuits for him one day. I thought I would surprise him. The biscuits came out as hard as rock. I hid the biscuits from him, and put them way up on the cabinet. When he came in from work, he slammed the door and down came the biscuits, as hard as rocks. I started crying. He said, "Don't cry; that's how you learn."

Andy at this time was working for the Automotive Warehouse. Things were now falling into place. It was getting little easier for me. Then I found out I was having my second child, Naydean Viola.

There was a difference of 13 months between Sonny Boy and Naydean. This time I knew what to expect. I was expecting the worst. God helped me: Naydean was the best baby ever. She didn't cry, not even when she was hungry. One day when I was busy doing my chores. The neighbor asked me if Sonny Boy could go outside and play with her little girl. She said she would be watching them. In the meantime, the children disappeared. She knocked at my door to ask if I had the children. I said, "No, you were supposed to be watching them." We looked all over for them and couldn't find them. I called my husband and he called the police. This was late in the evening. By that time my nerves were at edge.

The police said, "They must have fallen in the conservancy ditch. That was just around the bend. They were lost from ten o'clock till seven o'clock in the evening. They were playing in this neighbor lady's garage. She said she heard the dogs barking in the garage. She had vicious dogs. It's a wonder the dogs didn't kill them. She brought the two babies to us. The two babies were covered with white stuff. I thought they would be crying. They were unconcerned, wondering why these people were crying? They had no food for nine hours. This was a miracle for me. The Guardian Angel was looking after them for me.

Sgt. Andres N. Jaramillo at Fort Bliss, Texas .

In the meantime my husband was offered another job at the New Mexico Timber in Bernalillo, New Mexico. We lived in Bernalillo for about one year. When he got his calling for the service in WW II, he knew he would be drafted. He was only 21 at the time. I prayed hard so that he wouldn't pass, but he did. A day after he was called into service, our house burned down with everything in it. Faulty wiring caused the fire. The neighbors helped me take my babies out. I was stunned in the middle of the front room. I could not move. Our neighbor commented, "You're nothing but a baby yourself."

Naydean and Sonny Boy (Andrew) at Ft. Bliss, Texas.

We both got into the car with our babies. We were both crying. At that time no welfare. People were very generous, offering us a place to stay. They helped us with clothing and bedding. My sister bought the two babies clothing. We both didn't know which way to turn. We had each other and our babies. Material things can be replaced, so we comforted ourselves. We had our ups and downs in our young age.

My Husband passed his physical for the service. We left Bernalillo two months later. He had to report to Fort Bliss, Texas. It was hard on both of us when we had to be separated. My mom had moved to Los Lunas. I hated to impose on my mom again, and with two babies. I stayed with her three months. It took my husband that long to get living quarters. When he had finished his boot training, he got stationed in Fort Bliss.

Since he was a good typist and had a high score in math, he was placed in the Financial Department. We were both happy that he was stationed in Fort Bliss. At least something was going our way. We had a good life in the service. We lived close to the theater,

N.C.O. Club and commissary. One thing bothered me: people were so prejudiced in Texas. The people who couldn't speak English very well and the people with dark skin were humiliated.

It didn't take my husband long to make sergeant. We lived in Ft. Bliss for four years. My third child was born in Fort Bliss, at the Beaumont Hospital. I was treated great since my skin was white. They thought my husband was Greek. Andy's mother was Italian and father Hispanic. I guess that is why he looked Greek. The Blacks were in one ward and Latin origin in another.

Two weeks after Paulette Eunice was born, Andy was sent to Fort Sam Houston; I went back to Belen, New Mexico, where my dad had retired. My husband was discharged from Fort Sam Houston. He had ulcers.

We stayed in Belen, New Mexico, for a couple of months. Times were hard. He thought that his job would be waiting for him in Bernallilo, but it had been abolished. There were no jobs to be had in New Mexico. Rumors were that Barstow was hiring at the Marine Base. My husband and my brother who was just out of the Navy went to Barstow, California.

They were hired immediately. Living quarters were issued the next week. We came by train, my sister-in-law, my mother and I. The apartments were very nice. They were fully furnished with two or three bathrooms, depending on how big a family we had. We had a lot of conveniences at the Annex: a movie theater, swimming pool, and commissary. Lifeguards taught my children

The young Jaramillo Family at Daggett Annex, USMC Supply Center, Barstow, CA.L. to R. Geraldine, Mary Ruth, Pita, Paulette, Naydean & friend..

to swim, giving private lessons for anyone who was interested. I took advantage of anything that bettered my children in their adult lives.

I did not realize that there was segregation here. Each barrack was assigned by nationality. The Blacks in one, Hispanics in another, and Anglo-Americans. We had Mexican people living in our barracks. Their language was quite different than the people from New Mexico. I thought I had left discrimination back in Texas. My sister-in-law wasn't too happy—she went back home.

I put the prejudices aside, for the advantages. People in general were nice here. The lifeguards loved my children. I took them swimming every day during the summer months.

We had a wonderful chaplain, Father Fitzgerald. He gave Mass on Sunday. He also taught Sunday school teaching my Sonny Boy to be an Alter Server. Paulette Eunice was only nine months old and she learned how to swim.

Three of my children were born at Daggett Annex. Geraldine Ann Dolores, born at Wilkes Hospital in Barstow, Mary Ruth and Lupe Elma Corine at Dr. Graybill's.

We lived at the Annex for a period of seven years. The Annex was to be closed or torn down. We were given enough time to rent or buy a house. Mostly, all the people lived at the Annex had come from the military. So we used the G.I. Loan to buy a house. Anthony Joseph, was born in our new home. We were blessed again with our seventh child. I was not suppose to have another child, my blood pressure was sky-high. Dr. Graybill said, "There is not going to be a next time." So, he took matters into his own hands. He treated me like a daughter, not a patient.

We were lucky enough to get our home right next to Cameron School. I would walk the children to school every day and wait for them when school was dismissed.

As time progressed, the oldest children went to Barstow High, and the youngest to Kennedy High. Both of the boys were involved in sports. Andy, their Dad, loved all kinds of sports. Sonny Boy, my oldest, played catcher for the varsity. Both boys played football, both lettering. The boys and girls all played in the school band. I have a box full of their certificates for individual competition and band competition. We were always there to give them moral support.

My husband at this time had been promoted to Quarterman or General Foreman. The general as Head Council also appointed him for E.E.O. Committee. This involved a lot of his time. Counselors resolved most of the problems brought to them by employees and applicants, frequently because he was able to suggest corrective action, which was appropriate and practicable, but often because he was a knowledgeable, sympathetic person who diplomatically corrected misunderstandings. He would help forge stronger and more efficient relationships between human beings at the same work environment. Aside from his vital role to the processing of individual complaints of discrimination, the counselor functioned as the eyes and ears of the Commanding General.

My Husband was involved with Minor League East and Senior League. Our youngest son played first base. Many years we were involved with youngsters. I would fix sandwiches for the whole team. We'd take them to Dodger Stadium to see the Dodgers. All our hard work paid off. We won the championship. Danny Norman was one of our players, later playing in the major leagues.

My husband was involved with the school as Booster Club President for Kennedy High. He also attended many schools himself: Aberdeen, Maryland, quite often. The University of Southern California, Department of Industrial Engineering, San Diego, Parker College at the Marine Corps Center. My house is full of his certificates.

We also gave our girls dancing lessons, tap and ballet, baton lessons. My youngest daughter Lupe for the anniversary celebration of Barstow was chosen for PeeWee Queen. We had the girls in Brownie's, Girl Scouts and the boys in Boy Scouts.

My husband was an intelligent and aggressive man. He pushed education to his children. He would say, "I don't want my kids to have the hard knocks I had." He honored the flag of the United States. If he were to see the flag on television, he would stand up and salute it. When election time came, he was first in line. He said, "It's an honor and a privilege to vote."

Both boys worked on weekends. The money they earned

Lupe and Anthony at the Kelly Street house in Barstow.

was *saved* for them. This gave them an edge of responsibility.

They attended college. Sonny Boy attended San Bernardino Valley College. We rented him an apartment, but he would come every weekend so he could take the cream pies I had baked for him. I would make food for the week.

After two years my oldest son took matters into his own hands. He brought us the girl he wanted to marry. My husband and I were upset at first. Angelina Hurtado was her name. I wouldn't trade her for the world. He worked for the Bun Boy in his teenage years. He was hired at the Bun Boy as Manager. He furthered his education and became a New York Life Agent. He is also a chartered Life Under Writer, Chartered, Financial Consultant and Registered Investment Advisor. He is also a member of N.Y.L.T.C. University Leaders for Life Member of the Million Dollar Round Table. Top of the Table for 28 years. M.D.R.T. is recognized throughout the industry as the standard of excellence in Life Insurance Performance.

Anthony Joseph, we called him, Brother. He went to junior college here in Barstow, earning his Associate of Science degree.

From here he went to the University of Riverside and Bachelor of Science. He also would come home on the weekends for his homemade pies, and food for the week. He also worked in the weekends, helping my son-in-law at Purdy's Meats, and working for Vons.

Anthony Joseph worked as counselor for juveniles at Riverside. He found this work too dramatic. Now he is working for Albertsons as manager. He is credited for the opening many Albertsons stores. He made them flourish.

Naydean Viola, my oldest daughter, after graduated was hired at Contel. She was an instructor when she retired. She wasn't old enough in age, but she had enough years with Contel to retire. G.T.E. took over Contel. We are proud of our daughter for her accomplishments. She still is a busy bee, helping the town

Daggett School in 1950.

of Barstow. M. D. Williams was her husband, whom she lost at an early age.

An independent person, she depends on herself.

Paulette Eunice owns Purdy's Meat Market with her husband Tommy Purdy. They have made their business a big success. They too had the misfortune of their store burning. They picked up where they left off and bounced right up again. It was a lot hard work. *My family are survivors.*

Geraldine Ann Dolores works for the Marine Base. She's

another independent person. She works for Repair Division, the backbone of the Division. There is nothing this child could not do, from decorating a house to fixing computers. She is married to Eddie Simko, a private investigator.

Mary Ruth is my miracle child. She was in a bad car acci-

dent, her pelvis being shattered, her intestines paralyzed. She was bleeding internally, had a head concussion. I had her six months in a hospital bed. The doctor said she might never walk again. When I was getting her ready to take her to the doctor's office, she got up from the wheel chair and started walking. Mary Ruth and her husband Alfred Arguelles own a big alfalfa ranch in Newberry. They also own quite a few homes in Newberry and

Achievement Award plaque given to Andres N. Jaramillo for "Meritorious Civilian Service to the Marine Corps and this Base April 21, 1947 to November 21, 1977."

Barstow.

Lupe Elma Corine, the youngest of the girls, has been an aerobics instructor for quite a few years. My daughter is in ship-shape condition. Her husband is Mark Mauldin, a Barstow Police Officer, also in the SWAT team. She also has a degree in cosmetology.

My son Anthony is married to Kathy Fisker. Since she wanted to become a nurse, he put her through school and she is now a registered nurse.

Our lives were centered on our children. We are proud of all of them and God has blessed us with a wonderful family. I thank God everyday for giving us such a blessed family.

My husband's passing left us all empty inside. We miss Him so much. My husband was the foundation and the rock we stood on.

We were the pillar and my husband the rock. He held us all up with unconditional love and values.

The End of an Era

This is at the end of an era. Our children are all raised and settled with families of their own. We were always there for them when they needed advice or had problems. Now was our turn to think of ourselves a little. Things do not work out as you plan them. My husband was at work; he went to check on his men, although he had no business there. He had his foreman to do it for him. That day he fell from a five-foot plank and hit his head on the railroad track. He weighed two hundred and fifty pounds at that time. He really got a good wallop. It's a wonder he did not kill himself. He was away from work for a period of three months. They made him go back to work, but he was not feeling too good. There was nobody to take his place. He was an asset to the Marine Corps. He was 55 years old. He had completed his 30 years at the Marine Corps base. Finally, the Marine Corps let him go. He had to go out on a medical. His health was not good after the fall.

His hopes and dreams of traveling never left his mind. I had to give in to his every whim. He deserved it. He had worked for all of us. Andy loved life and beautiful things. We made our first trip to the Caribbean. We visited Puerto Rico, St. Thomas, the Rain Forests and the Dominican Republic. The name of the ship was the *Song of Norway*. I could have stayed there forever. Later we visited Hawaii, Maui, and all the surrounding islands. We went to Mexico City, which I did not like much because they lived in a fast pace. The cars would go at a hundred miles an hour with no respect for pedestrians. We wanted to visit the place where the Virgin had appeared to the Indian boy Juan Diego. The surrounding towns like Taxco were nice to see. They lived in medieval times, lots more history. We made a trip to Spain, visited Madrid and Toledo. In Portugal, we visited the Lady of Lourdes where she appeared to the three children. Morocco was unique—the people kept their customs and cultures. They ride camels and keep veils over their faces. The Casbah was historical. Our trip to Canada was great, beautiful lakes, beautiful scenery. The glaciers were stupendous. Our last long trip was on the train, a trip around the USA. It took us 26 days—the most beautiful country in the world.

Sometimes, I go back in time and reminisce when our children were small. On Christmas Eve, he took them all for a long ride, waiting for Santa to come. I would put presents under the tree, while they were gone. He would come an hour later and say, "I think Santa Claus has been here."

On Halloween, he decorated the outside of the house with paper bags and a candle in them. It was decorative.

When they were little, there was not an animal he didn't bring home to them. They had rabbits, lamb, cats, dogs, turtles, and a chicken, which turned out to be a rooster. The kids named the rooster Henry. He was the meanest rooster. He ran after the neighbors and pecked them. We finally had to get rid of him. I remember for Easter, he brought the children, a small rabbit. Naydean was walking in the door and stepped on him. She had not seen the rabbit run under feet. The other kids were crying calling her, "*You* murderer you." So, the next day, he told them to go outside to see if the rabbit was still in the cage. He had told them he had taken the small rabbit to the doctor to make him well. They found a rabbit, a huge rabbit. The children asked him, "How come he grew so big?"

Now, believe me, he was also very strict, when he gave a command they had to obey. He was strict about their getting good grades in school. He would not help them with homework. I had to help them all and my grandchildren asked me, "How did you do it, Meme? I would have left them [my children] on somebody's door step."

I'm always bragging about my children but they are worth it. For our 50th anniversary they made a big celebration for us—a beautiful Mass with the mariachi band. We had a big reception at Rositas. It was such a beautiful day it will live in my mind forever.

I thank God he gave him a boost to accomplish his last wishes. I could tell he wasn't feeling very well. He wanted to visit every nook and corner of the world. I could tell he was feeling a lot of anxiety. He wanted to go Albuquerque, New Mexico, in the truck, to get some green chili. My sister and brother lived in Albuquerque, and he also had relatives. He said, "We can visit them, and bring chili on the way back." Our children were opposed; they did not want him driving that long distance. They had noticed a change in him. Paulette and Tommy, especially, noticed a change. He was helping them at the store, banking for them.

All our trips were usually taken in September. We made it a point to be home in December. We had our family gatherings on Christmas Eve. We cleaned out the garage to make room for the table and chairs. Our family had over-populated the house. We made preparations a week before Christmas Eve. We made arrangements for the girls to come on Saturday. Nobody worked on Saturday, so there would be no excuse. They all had to pitch in and help. We made twelve dozen tamales. Andy already had the masa or dough mixed and the chili made. I helped him, of course. The girls' job was to spread the masa or dough on the corn husks. My husband did the fillings. We also made hominy, with pork meat and chili pods. I gave the girls the recipe for *bizcochitos*, our favorite cookies. The girls made all the goodies. These cookies were originated from Mew Mexico. We learned to make tamales here in California. I don't recall having tamales in New Mexico. I have quite a few recipe books from New Mexico.

I think the origin of different foods, culture and customs is a follow up from your ancestors. My husband's mother was Italian and father Hispanic. His first language was Spanish. I, on the other hand, my first language was English. My grandmother was Irish-that's on my mother's side. Her father's name was William Riley Milligan. In comparison to my husband and me, the foods, customs and culture were somewhat different. It could be that the different food, customs and cultures are derived from the heritage of the grandparents who taught our siblings.

Well, we made the trip and visited our relatives. We brought chili for the family. The next day was Sunday. We went to Mass and came home. It was a day I don't like to remember. Andy had a seizure that lasted one hour and thirty minutes. Paul Purdy, my grandson, gave him C.P.R. The paramedics rushed him to the hospital. The doctor said the seizure had damaged him because of its severity. From that time on his health was failing. The only time his eyes would light up was when his girls and boys came in the door.

There is nothing more intimidating than to see a man of his caliber struck down with such a ferocious fury. He broke our hearts. We miss him so much. He left us with an emptiness that cannot be explained. He left us, with his undying love and values.

Two lives entwined as one
Our commitment in life was
To God and our love for each other
Our love brought us through the
Trials of life.

Our love will never die.
We were blessed with seven children,
Who, carries on our love and heritage.

Memories of Dad
By Andrew J. Jarmillo

I was born May 28,1940, at a time when my parents were living in New Mexico. I have no recollection of my early years in New Mexico. They say that our earliest memories usually involve a trauma and mine does. I was only two or three years old and my father was in the army and stationed at Fort Bliss in El Paso, Texas. I had contracted scarlet fever and was quarantined with my mother. She gave me a lot of loving care for about six months. My dad had to stay with my sister Naydean, who was still a baby at the time.

When I was four we moved to Daggett Annex, Marine Corps Base, which had both Marine and civilians living on base. My keenest memory of living on the base with my mom and dad and five sisters was being the center of attention and seeing how much they loved and cared about me.

The dominant figure in my youth was my father, a man about five foot nine inches tall, but he looked ten feet tall to me.

He worked hard to support his family. I remember he always had a newspaper tucked under his arm. He was always whistling and stopped to greet everybody. He loved life. My father was self-confident and totally secure. He had inner strength and the will to endure anything life threw his way. He portrayed himself with the "macho attitude" but he was very sensitive to how others felt. He was always able to be both mighty and sweet. My dad was truly my hero.

My mother was the eldest of her generation of four children. My mother had a high school education and graduated with top honors. My grandmother had named her Guadalupe, but my mother was known as "Pita" to all who knew her.

When I picture Mom, she is always working around the house, always in motion chasing two boys and five girls, cooking, cleaning, washing, ironing and sewing. Mom was a staunch democratic supporter and so was my dad. My dad was the eternal optimist and my mother the perennial worrier.

I had a great childhood, teenage years and wonderful loving parents. I am proud to be their son.

by Naydean Jaramillo Williams

As I look back at my life today, I feel privileged to have been taught characteristics, such family unity, responsibility, good *work* ethics, compassion, obedience, gratefulness, sincerity, forgiveness, truthfulness, and the love of God. Through life experiences those character issues were how I made life-building choices, thanks to my mom and dad.

When I first came to Barstow, actually Daggett Annex, I was about three years old in the year 1947. My father was hired at the Marine Base, which was a wonderful opportunity for him to support our family. Jobs were hard to get at that time. Daggett Annex was a housing location where we first lived. There were many buildings or what they referred to as barracks. The barracks were numbered and had apartments with up and down stairs. Military families and civilians were assigned where to live according to rank and color, which I thought was odd.

We had a grade school; everyone was in the same classroom no matter what grade we were in or what age. I remember my first teacher Mr. Emil Kline, a kind and caring teacher. What stands out most in my mind was his kind way he treated the students. As time went on, that old classroom closed and the town of Daggett built a new school, and we were bussed to school. William Venner was a wonderful teacher I had in the 5th grade.

My dad, brother and I delivered newspapers in the early morning and in the evening time. It was a wonderful time for me, learning how to collect money, remembering numbers, and talking with different people. We also had the opportunity to have all the privileges that the service people had like using the swimming pool, theater and the commissary.

Family size: When we first arrived in this part of the desert, my parents had three children, Andy, Naydean and Paulette in that order. While we were at the Daggett Annex, the size of my family increased with the birth of my three sisters Geraldine, MaryRuth, and Lupe. When the housing facilities at Daggett Annex Base closed, we had to move to Barstow. There was another birth, my brother Anthony, altogether a grand total of seven children.

We moved to 929 Kelly Drive where my mother still resides today. Can you imagine the fun of seven children plus my parents living in a three-bedroom, one-bathroom house? In one bedroom there were two beds, one full and one twin, three of us sleeping in the full bed, and the other two in the twin bed. My older brother had the smaller bedroom to himself. My little brother was in my parents' bedroom in his baby bed until he was of age to join the rest of us. We were crowded, and didn't even know it, because we had lots of love.

Even with all seven children my father and mother kept us involved in community and school functions. I took dance lessons and performed for many functions in town and at Fort Irwin. My baby sister Lupe also danced and performed for community functions. Lupe was the first peewee queen for the City of Barstow for the May festival. My father was always involved with youth baseball; in fact his team was champions for the City of Barstow. In school functions my father was president for the Parent Teachers Association.

We all graduated from Barstow High School except for Anthony who graduated from Kennedy High School. Both my brothers attended and graduated from college. Andy, my eldest brother, married Angie and is a New York Life Agent doing well. My youngest brother is married to Kathy and works for Albertsons as Meat Manager. I was married to M. D. Williams (now deceased) and worked for the telephone company (Contel), starting as a telephone operator, promoted to a Training Specialist and retired after 34 years. Paulette married Tom Purdy, and they are the owners of Purdy's Meat Company. Lupe married Mark Mauldin, who works for the city of Barstow as a police officer, and she is an aerobic teacher. Geraldine is married to Eddie Simko and works at the Marine Base. Mary is married to Alfredo Arguelles; they have a ranch in Newberry, eighteen miles out of Barstow.

A special event in my family was Christmas; Dad called all of us girls together a couple of days before Christmas Eve to make tamales. We had so much fun. My dad loved his girls. He would be in a public place of business and if any of us were around, he would say to anyone who was there, "Don't you think my daughter or daughters are beautiful?" We wanted to just hide under a table. When we were invited to any of our relative's weddings, of course we would all go. He would dance with all five of us and with my mother without getting tired! What a man! My parents instilled the importance of family values along with being a productive citizen, which we have carried through to our families. My dad was a very proud man and proud of his heritage, which he passed on to his family: We all know who we are and what we stand for.

The Jaramillo family is a hard working, civic-minded, compassionate family. We take pride in our children's achievements and in each other. Our faith in God, love of country, pride, and the love of our family, is the substance of our inner being.

by Paulette Jaramillo Purdy

What makes wonderful parents? People who are capable of giving all, whether it is physically, morally or mentally—that describes my wonderful parents.

Dad grew up in an era having to face adversity and prejudice. He overcame all, through education, pride, honesty, and the determination to work hard to succeed. These are the values with which he used to raise his family, with Mom at his side. We were raised in a home of love, honesty, trust and respect for others as well as ourselves. He taught us that there was no shame in failure, but only in not getting back up and trying again. "I can't" or "I don't want to" were not acceptable. By doing the best we could and the knowledge of his being proud of our effort were really the only rewards needed. The encouragement was always there, and believe me, Mom and Dad were always our biggest cheerleaders whether it was playing in the school band, dancing in a recital or playing a sport. Dad made sure we all played an instrument. And there was none of this stuff of starting something and quitting in the middle. If you started it, you finished it.

Education was an important part of our lives. I remember Dad saying, "If you're too sick for school, you're too sick for anything else." If you missed school, you had better really be sick. My parents never missed one parent teacher conference for anyone of us from kindergarten through high school. And our grades had better be up to his standards too! This was his way of making sure that we would be able to function and succeed in the world once we were on our own. He was there as a safety net, but he expected us to stand on our own two feet. He wanted so much more for us than he had when he as growing up, and he knew education was one sure way of guaranteeing that.

I have so many wonderful memories of my father. He loved to dance. When we went to wedding dances he would wear my mom out and then all five of us girls. I don't think he ever got tired. We girls were happy when the granddaughters and great granddaughters came along 'cause that just meant he never lacked a dance partner. My dad came into Purdy's Meats every day to see us, most of the time he would have lip stick on his face from mom. Tommy would always ask him if he been chasing Mom around the house again. Daddy got a big kick out of this.

Our holidays were another wonderful memory for me: Thanksgiving meant giving thanks to God for his Blessings to turkey and dressing with red chili. Christmas was another special time we celebrated with *pasole*, tamales and Christmas Mass. Family togetherness was always important during these times. These traditions we still use to this day. I remember, as a little girl, going on a field trips to San Diego. He made this trip so important for me: making sure I had a special lunch and enough money so I wouldn't feel out of place.

As a grandfather he measured up as the best, attending all their activities and being their biggest cheerleader. The tradition continues. No matter where he was or even if he knew the person next to him, he would always say, "Look at my girls, aren't they beautiful?" What a wonderful father and grandfather.

I wrote this poem for my dad:

Today I Say I Love You
by Paulette Jaramillo Purdy

Tomorrow is too late
To hold your hand and touch your face
For in my heart no one can take your place.
Tomorrow is too late, for life ends for all,
that is our father.
Good-bye for now Daddy—I love you.
Thank you for your guidance, your love, and your sacrifices and for being my father.
You are the best example our family has.

Touching Everybody He Met
By Geraldine Jaramillo Simko

To my daddy, family was a priority. He instilled this in every one of his children, grandchildren and great grandchildren. Out of my parents' seven children, I was the middle child, born in Barstow, California, at Wilkes Hospital. As the middle child I never felt neglected. Mom and Daddy made sure each every one of us felt special in our own way and that we could become anything we wanted.

My mother told me when she was potty training me, I refused to go; I wanted Daddy to take me potty. My daddy had to take off work to come home because my stomach was getting so big. That is how dedicated my daddy was to his family.

Daddy had a character that could make you laugh when you felt down, but also received a tremendous amount of respect. He touched everyone he met. Everyone knew "Andy."

When I married my husband, Daddy told my husband he was now part of the family. My husband has always felt that my daddy was his daddy too. To this day, Daddy made such an impact on my four children that I see a piece of him in all four children and they're passing this teaching down to their children. Daddy was so thick with knowledge, faith and life experience—that's what made his advice so genuine.

When we lost Daddy on September 19, 1999, it felt as if my life was also ending. The person who played a huge part in my life had left this world. I felt like a lost little girl. I came to realize, Daddy is still with me, in my heart; always I have the promise of God that I will see him again, healthy. My job on earth is not done, but when my time is up, Daddy will be there to greet me.

As Seen by Lupe Mauldin

My name is Lupe Mauldin, my maiden name Jaramillo, the second youngest of seven children of Andy and Pita Jaramillo. I was born in Barstow, California, but my heritage is from New Mexico. I can still remember vividly many visits my grandparents and other family members made to Barstow from New Mexico, and the culture they spoke of and brought with them. For instance every year my father would either go to New Mexico to buy green chilies or make arrangements to buy some from someone who would be making a trip. Once the chilies arrived all of the family helped roast and peel the chilies. Then they would be bagged and placed in the freezer to be used throughout the year. Growing up my family would have beans, tortillas and chilies almost every meal. I did not realize until I was older and started school that not everyone had tortillas and beans regularly. Another tradition we always enjoyed was making tamales for Christmas. Dad cooked the red chilies, beans and pozole and all of our families got together and helped make the tamales.

I remember when I was dating my husband; I was embarrassed to have him come over for dinner because I thought he would not like our culture and food. Finally I had him over and he loved the tortillas, beans, chilies and our culture. He liked us for being us, and loved me for being me.

My father talked to us often about the depression in New Mexico and how he would work in the pinto bean fields. After the harvest the owners allowed the families to come behind and pick-up the left over beans. Dad would say, "Had it not been for the pinto beans his family and many other families from New Mexico would not have survived!" This was a story he repeated often to us in his last days with us.

As I was getting older I realized that many people tend to place all Latino's in one box. I soon realized that there is a vast difference among the Latino community and that our New Mexico heritage has its own unique culture, food and language.

My family, including my husband, daughter and grandchildren, has all benefited from the New Mexico culture that is part of my heritage and will be until the day I die.

I have included dad's *pozole* recipe just in case you decide you might want to experience a little of the New Mexico heritage, Jaramillo style!

Pozole Recipe:

2 cups lime hominy
4 dried chile pods
2 tsp. oregano
2 chopped garlic cloves
6 cups water
2 tsp. salt
1 lb. pork (cut in sm. squares)

Rinse the lime hominy 4 or 5 times, put water in and boil until hominy breaks open. Then add seasoning and cut-up pork squares. Cook for ½- ¾ hours. Serve with warm tortillas.

For quick recipe, buy canned white hominy, dump in pot with pork and seasoning. Cook until pork is done.

Bizcochitos Recipe

1 cup of water
2 cups of shortening
1 teaspoon of anise
2 eggs
6 cups of flour
3 teaspoons of baking powder
1 teaspoon of salt
¼ cup of water

You can substitute the ¼ cup of water with red wine or milk. Cream or whip the sugar and shortening, add two eggs and the teaspoon of anise. Add the six cups of sifted flour, three teaspoons of baking powder, one teaspoon of salt, ¼ cup of water, wine or milk.

You have to knead the dough, form into a ball, roll out and use a round cookie cutter.

Mix sugar and cinnamon in a plate or bowl, dab each cookie with cinnamon and sugar. Be sure the cookie with cinnamon and sugar this side up. Place in cookie sheet and bake.

We Know Who We Are
by MaryRuth Arguellas

The guiding light of my life, were with my dad and mom, Andy and Pita. They protected us through our childhood, and taught us right from wrong. I realize then that my dad knew who he was, and was proud. He was just for us. He was a man who loved us.

The one thing that stands out in my life is that I really didn't know my heritage. What I do remember is that we lived in an all white neighborhood. All my friends were white, until I started high school. That's where I found out I wasn't white. I was Hispanic. The majority of the Hispanics hated me because; they said I thought I was white.

I knew I was proud of who I was as a person, but I just didn't know my heritage.

I felt my parents went through a lot with prejudicial issues to keep us kids from getting hurt. That is the way they protected us. Now that I am a parent, I'm married to a Cuban who has

Andres and Pita's 50th Wedding Anniversary: L to R: Geraldine, Paulette, Naynean, Pita, Andrew Jude (Sonny Boy), Mary Ruth, Lupe and Anthony..

taught me how to celebrate who I am. He helped my children to speak in their native language and to be proud, of who they are.

When my father became ill, I learned much more about my heritage. As I went back into the past, my father's eyes would light up when he would talk about his family in New Mexico. My father accomplished a lot in his life: Becoming the first Hispanic general foreman at the Marine Corps Base, and became E.E.O. Council for minorities. I also worked at the Marine Corps for a period of ten years. I was proud of my father. People would stop at work and ask me, "You're Andy's daughter, aren't you?" This made me feel proud of being my father's daughter. People who are prejudice are not well educated. We are all created equal. You know he was a wonderful man and really helped the minority people at the Marine Base and in life.

Dad had made a better world for us. I realize then that my dad knew who he was, and was proud. He was just for us. He was a man who loved us very much.

It's what you make of yourself that matters. And in the process of making the world better for us, he also made the community a better place.

Part 2
The Olguin Family of the Mojave Desert

Jesus Olguin and family in1925 in New Mexico. Sitting, Jesus Olguin and Cleofus Olguin (Gillie's grandparents). In lap, Dennis Olguin (Gillie's oldest brother). Standing, Nicholas Olguin (Gillie's uncle), Manuelita Olguin (Gillie's mom), Jose David Olguin (Gillie's dad), Eutemia Olguin (aunt), Maria San Juan Olguin (aunt).

The Olguins: Our Story

The American Dream

Barstow has been the home of the Olguin family for more than 50 years. In fact, Barstow and the Olguins have become almost synonymous with each other. As some might say, you can't think of one without the other. Although we had brief stints living in New Mexico for varying lengths of time, we will always call Barstow our home, because most of us were born there, went to school there and in some cases are still living there today.

Gillie and Vina's Wedding in, Nov. 12, 1949 in New Mexico.

For many people like Gillie and Vina Olguin, Barstow became a symbol of living the "American Dream." This small, quaint town provided the opportunity for Gillie and Vina to carve out a living, raise a family and become pillars of the community. From humble beginnings, these two simple people created a family that currently includes eight children, sixteen grandchildren, and one great grandchild. And everyone still has favorable ties to Barstow.

Of course living the American Dream did not come without a price.

In 1950, at 21 years of age, and with 10 months of marriage under their belt, Gillie asked his new bride, Vina, to leave Belen, New Mexico, and move to California to start a new life together. This was a very difficult decision for Vina because she had never been more than 60 miles away from home or spent

Gillie and Vina's 40th Wedding Anniversary in 1989, Barstow.

any significant time away from her large, close-knit family. Thus, this was a difficult decision to make, coming with much anguish. However, Vina knew that despite her heart telling her not to go, and her head told her their future would be better served in the land of opportunity called California and this small military town of Barstow.

Why Barstow?

The majority of people who grew up in the small towns of New Mexico, like Gillie and Vina, typically stayed close to home to live and raise their children. However, in August of 1950, Gillie and Vina ventured out to California to vacation with Vina's brother, Daniel, who was living in Barstow and working at the Marine Corps Supply Depot (now the Marine Corps Logistics Base).

During the visit, Daniel asked Gillie to accompany him as he went to pick up his paycheck at the Marine Base. While sitting in the truck waiting, Gillie was confronted by Major Elrod, who asked, "Are you Gillie Olguin?" Gillie replied that he was, and the major handed him a form and said, "Take this form to sickbay to take a physical so that you can report to work tonight." Shocked, Gillie didn't know what to think. Just then, Danny returned to the truck and said, "So, now you have a job!" Ultimately Gillie turned his two-week vacation into a 50-year stint in Barstow!

The Early Days

Initially, Vina had a tremendous amount of trepidation about living in Barstow. Each day she struggled with her daily routine, because she missed home, missed her parents and missed her family. Additionally, she was exposed to racism for the first time, something she never expected, nor was she prepared for. After a few months of getting acclimated to Barstow, Gillie and

Vina applied for and were accepted to live in government-owned housing. There were 20 buildings with 9 apartments in each. Fifteen of the buildings were for Anglos, two buildings were for Hispanics, two others were for African-Americans and one was for Native Americans. Unfortunately, because of the unusual spelling and pronunciation of Olguin, Gillie's last name, Gillie and Vina were incorrectly assigned to the Anglo buildings. This just made matters worse for Vina and every day Gillie would return home to find Vina in tears, crying, "Why did they put us here? Why can't we be with the other Hispanics?"

Gillie and Vina in front of the government housing in Daggett in 1950.

Gillie responded, "Unfortunately, there is nothing that can be done and we are lucky to get this furnished one-bedroom apartment for only $27 per month."

To pass the time, Vina cooked all day in preparation for Gillie's return from work. Her specialty was tortillas, a New Mexico specialty, which over the years became the family favorite. Once she began making the tortillas, the aromatic smell would fill all of the neighbor's apartments. This came to be quite a good omen for Gillie and Vina.

One day as Vina was making her famous tortillas, a curious neighbor knocked on the door because she could smell something good, and just had to know what it was. When Vina answered the door she told the neighbor that she was making tortillas and invited her inside to try one. After Vina demonstrated the proper eating technique, the curious neighbor was hooked. Next thing you know, each night all of the neighbors were stopping by each to sample Vina's culinary delights. This was the beginning of acceptance for Vina.

Though she always believed that they would move back to New Mexico, saying things like "someday we'll move back home," even convincing Gillie to move to Roswell, New Mexico, for a short time. But Barstow was quickly becoming their home.

However, Gillie always tried to make Vina happy, and Vina always had a yearning go back to New Mexico. So, one day, while visiting Vina's family in New Mexico in 1962, one of Vina's brothers told Gillie, "I've heard that they are hiring at Walker Air Force Base in Roswell, why don't you go and see?" Unfortunately, Gillie was unable to go because it was late on Friday evening and he had to be back at work in Barstow on Monday morning. So Gillie called on the phone, and they hired him without an interview!

The transition was difficult; times were tough, as Gillie took a pay cut from $3.05 an hour in Barstow to $2.25 in Rosewell. However, everyone tried to help. One time, after getting settled in their new home, they went to visit another of Vina's brothers, Zeke, who owned over four thousand acres of land. Zeke was a great guy who was always trying to help others. "Do you have any meat in your freezer?" he would ask. When Gillie replied that he did not. Zeke said, "How would you like to take home a little goat?" The first time, Gillie, said "great," not knowing exactly what Zeke meant. That night, they got in Zeke's truck and went to the mountains, where they drove slowly through the brush with a bright flashlight. Two lights lit up, which were the eyes of a deer. They shot the deer, illegally. They had meat all year long.

However, that year in Roswell was tough financially, and Vina realized that the family would have a much better life in California. Gillie and Vina sold their house and moved back to Barstow to begin raising their family. And once they got the family started, they kept right on going!

Gillie's Story

Gillie was born on April 8, 1929, in Belen, New Mexico. His family ancestry has been traced back to 1560 when Juan Lopez Holguin was born in Fuente Ovejuna, Estremadura, Spain. Juan Lopez Olguin (note the change in spelling) and his wife were part of a Spanish contingent that settled in New Mexico in 1598, which was 10 years *before* the well-known Jamestown, Virginia, settlement of American history. Gillie Olguin is the 12th generation following Juan Lopez Olguin, and all twelve generations made their mark on New Mexico history. Gillie was the first to make the decision to move to another state when he and Vina relocated to Barstow, California.

Gillie was one of six kids born to Jose and Manuelita Olguin. When Gillie was about five years old, his mom and dad got food poisoning from a restaurant and Gillie was taken to his grandparents to live for a short time while his parents recovered in the hospital. That short time turned out to be seven years, which was quite common in those days as family members helped raise other family members' kids. Gillie loved living on his grandparents' ranch with the horses, cows, chickens, turkeys, and pigs and vegetable fields.

The ranch is where Gillie learned the true meaning of hard work and where he also gained his love for horses. He went to school at St. Mary's Catholic School and then to Belen High School. He made it to the 12th grade but never graduated, "cause his dad fell on hard times" and Gillie quit school to get a job and help out. That was in 1947. In 1949 he met Vina Chavez and that was

Gillie's single days in Belen, NM, 1949.

the beginning of a long and loving relationship that lasted until 1999 when Vina passed away from Alzheimer's Disease.

Although Vina passed away in 1999, Gillie continues to lives in Barstow. Alone at age 72, he has a couple of horses that he rides regularly, which keeps him young and vibrant. People say, "Gillie, aren't you a little too old to ride horses?" and Gillie responds, "Old is only in your mind, I'm not going to sit back in a rocking chair, I'm going to sit on a rocking horse!"

Vina' Story

Vina Susana Olguin was born in Duran, New Mexico, on August 11, 1928. She was one of three girls, in a family of 13 kids, born from Juan B. Chavez and Aurelia Chavez. As a teenager and the only girl still at home, Vina bore the responsibility for caring for her ailing mother and father, which meant she had to drop out of school in the seventh grade, and she never returned.

Since her niece Jane, who was her best friend, passed her in school the year Vina took off, Vina decided not to return. In 1948, Vina moved from Duran to Belen to live with her sister Concha.

Shortly after that, Vina's mom and dad also moved to Belen. Her dad was a rancher, but after they moved to Belen he went to work for the railroad. At this time Vina started working at Gilbert's Grocery Store. Vina was making $3 a day. However, through her strong will and determination, she saved up enough money to purchase a new top-of-the-line stove for her mother and father for $125.00. Vina also become a *mini-bank* for her brothers who were always

Vina's single days in 1949, Belen, NM.

coming to her for a small loans, most of which were never repaid and it wasn't expected.

During the time that Vina and Gillie were dating, Vina's brother, Benny, was away in the army. Benny left his car at home in Vina's care and told her to drive it no more than once a week. The car was a beautiful, bright red 1941 Buick Roadmaster convertible with a white top. When it was time for her to take it out for the weekly drive, she picked up Gillie, which was uncommon in those days (girls just didn't pick guys up guys for a date), and they would cruise around town with the top down looking like the richest kids in town!

They also had another form of transportation. Gillie used to ride a motorcycle and Vina rode on back with him. One day she asked if she could ride the motorcycle by herself. After much pestering and negotiating, Gillie relented. Boy, was he surprised at how well she could ride the motorcycle with after just a quick lesson.

How Mom and Dad Met

One day in February 1949, Gillie went to a drugstore where all the kids hung out, which included a particular a girl that he had a crush on. When he went in to the drugstore, Gillie sat next to this girl hoping to have the courage to ask her for a date. Also sitting in the group was another girl named Vina. As it turned out, the girl that Gillie liked was waiting for her date that was running late, and Gillie was disappointed. When the other guy finally arrived, they all walked outside together to talk. They all talked for a while and the couple invited Vina and Gillie to go for a ride together, but since Gillie didn't want to leave his car at the drugstore, Vina jumped in the car with Gillie and suggested that he leave the car at her house. He accepted her offer, and then the four drove off for a long ride to Mountaineer, New Mexico.

Vina and Gillie got to know each other on that ride, and Gillie thought she was wonderful. After that first fabulous day together, Vina agreed to see Gillie again.

At the time, Gillie was working at a bakery, delivering bread and pastries to local restaurants and stores. Vina worked in one of these small grocery stores, which became of particular interest to Gillie, since he serviced this store as though it was one of the larger accounts. He really wasn't interested in delivering extra customer service, he was interested in the beautiful girl who worked there! Gillie and Vina were married on November 12, 1949, and she was a beautiful and loving wife for years for nearly 50 years. As they say, the rest is history!

Vina and Gillie just married in 1950 in Belen, NM.

Eight Is Enough!

Most people perceive growing up in a large family to be difficult. That wasn't the case for us—the eight kids of the Olguin family. We all actually liked each other and enjoyed the large family camaraderie. Through the years, even though we had our challenging moments where we thought we hated each other, we came to appreciate the unique bond we shared from our childhood experiences and from our parents' guidance and leadership.

Elaine is the oldest of the eight kids. She relished her role as the pillar that the rest of us have all leaned on at one time or another while growing up. Elaine bore the greatest burden of us kids, as Mom and Dad were very strict with Elaine and did not allow her much freedom. We are not sure if this strict approach contributed to her getting married at 18 or not, but it definitely helped pave the way for rest of us of the kids. I remember that it was "huge" deal when Mom and Dad allowed Elaine and her best friend, Carmela, to spend a weekend with our Uncle Andy in San Francisco. We were all so envious.

When Elaine was a senior in high school Dad bought her a bright red Pontiac LeMans. But it was not given to her outright. She had to work for it. Elaine had to take over Dad's job as the janitor for the Department of Motor Vehicles. This was a part-time job that Dad, with help from the boys, had for over 10 years. Mom and Dad were always trying to show us that hard work would get you through in life, and that nothing is for free.

As the oldest boy, Gary was blessed with rugged determination but he was also born with cerebral palsy. This handicap became a challenge that Gary would use as a source to draw strength from throughout his life. Whenever someone told Gary that he couldn't do something because of his disability that just made him work even harder.

Gary loves to dance! He is also known as always being the brother who keeps our family history alive. He placed a keen sense of importance in communicating and building strong relationships with our older relatives. Gary has a great love for our relatives and a good ear for listening to their stories. Gary's fierce determination has motivated him to go back to college to get his degree at the age of 43, so that he could start a new career as an elementary school teacher.

Lori was the quiet, beautiful daughter. She didn't say much within the family, and she spent most of her high school days, with her bedroom door closed and the telephone permanently attached to her ear! It seemed like she came out for some occasional fresh air and for lunch and dinner. This was way before telephones had "call-waiting", so we could wait for hours for her to get off the phone.

Additionally, Ruben and Lori used to have some good knock-down-drag-out-fights over her corduroy Levi's because Ruben would wear her pants without asking and returned them dirty. Lori had a part-time job at Del Taco, where she was the patio queen, responsible for making sure that all the tables were clean. During high school, Lori was voted Los Amigos Princess at their annual ball. We were all quite proud of her!

Ruben was "Mr. Creative." The boys slept in a large basement bedroom and Ruben re-created the internal design many times. They went through a black-light poster phase, and then a ceramic head phase, which were based on art projects that Ruben created in high school art class. He also assumed responsibility for assembling beer pull-tabs from aluminum cans into long chains and hanging them throughout the basement bedroom to make it look like a jungle. As a kid Ruben wasn't much of a test taker and would get nervous every time he had to take one; therefore his third grade teacher, Sister Dulce Maria, gave him a piece of candy before every test and told him this was "courage" for him to help him take the test. Ruben was always socially outgoing, the first of us kids to be really social. He had many friends and was always headed to different places, like the beaches in San Diego, concerts in Las Vegas, or water skiing at the Colorado River. Usually, he always dragged along his younger brother, Michael.

Michael lived and breathed sports. His every waking moment was either playing sports or watching sports on television. His favorite sport was basketball, playing it every year from third grade through high school. From first grade on Mom always combed Michael's hair with *Dippity Doo* hair gel. One time when one of the nuns from school commented on how nice Michael's hair looked after recess, Michael commented that his Mom combed his hair with "goop," which was the family name for the stuff. The teachers all got a chuckle out of that!

Michael was the first of the Olguin kids to really put Mom and Dad to the test. Michael had a devious nature to him that must have forced Mom and Dad to ask themselves why they didn't stop after four kids. Whether he was sneaking out windows, coasting the car out of the back driveway, coming in after curfew, stealing letters from the drive-in marquee, or shooting a fire extinguisher at kids while they were trick-or-treating, Michael found unique ways to get in trouble. His antics must have added a few gray hairs to Mom and Dad's heads!

Natalie was a social butterfly in grade school and high school. She spent her grade school days in drill team and Mexican dancing, and her high school days cheerleading and hanging out with her friends, who we all referred to as the LGS (or Little Girls Society). She continues to spend a significant amount of time with those same friends today.

Natalie was also held the record for most nicknames in the family. Since she had a bad reputation for being a tattletale, she was thus called "Motormouth", "Nattletale", "SNAG" or just plain old "Mouth". As she has grown up, Natalie became probably more like our Mom than any of the other girls in the family. She's very nurturing and always making sure that everyone is cared for. Additionally, she has a keen desire to make sure that many of the family traditions and rituals that were developed over the years within the family are maintained.

Growing up, John was the best athlete and the best student in the family. Though Mom and Dad really encouraged us to study hard and make good grades, John went above and beyond what was expected. He worked hard in everything he did and made all of us very proud.

The fact that John survived at all is hard to believe, if you knew how "much abuse" that John suffered through as a kid, from his brother, Michael, who used to make John cry on a daily basis. Yet, John just took it all in and grew up to be the pride of the family. John has become the keeper of worthless tidbits of

Being silly and playing dress up, Mike, John, Ruben and Natalie in Barstow.

information. If you want to know who starred in some random sitcom in the early 1970s, John will know. If you want to hear a forgotten line from *The Godfather*, John will know. If you want to know some forgotten sports statistic, John will know.

Every family has to have a baby of the family. For the Olguins it was DeAnne. DeAnne spent her grade school days being tormented by John. Seems like John shifted his teasing focus from his older sister, Natalie, to his younger sister, DeAnne. If dealing with the teasing is supposed to make you a better person, then Natalie and DeAnne got a lot of early training!

DeAnne had an outstanding ability for spelling. She won numerous school contests and actually went to the County Spelling Bee. She was also very social and outgoing. DeAnne was on the drill team in grade school, and she was a cheerleader all through middle school and high school.

Artistic talent was also one of DeAnne's many abilities. She was always drawing and decorating. As the "baby" with seven older brothers and sisters, DeAnne was always getting plenty of advice and "help" with her many projects. Since DeAnne was just slightly older than some of her nieces and nephews, she was like a "big sister" to many of them. They often tried to emulate DeAnne, especially in the areas of artwork and creativity.

Outgrowing Our House

We obviously outgrew a number of homes as we added a new sibling every couple of years. We kept moving. Each time we moved to another house with another bedroom or a bigger family room or a bigger yard. By the time DeAnne was on the way, Dad realized that our the small three-bedroom house on Nancy Street was not big enough, so in 1968 he subsequently purchased a house at 121 W. Fredricks that ultimately would become "home" for all of us.

The Fredricks home, which features a huge basement, became the bedroom for all four boys. It's amazing how things change so quickly, because just five years earlier we moved to Nancy Street thinking that this huge house was going to be our "last move."

We remember Mom and Dad putting in striped wall-to-wall carpeting throughout the Nancy St. home, which we thought was pretty cool. The best part of the house was that only a mere five-foot wall separated our back yard from Fogelsong Park, which became our family playground away from home. We spent our days swinging on the swing sets and swimming in the pool, and our nights playing hide-and-seek.

Our Fredricks home was just one block from the high school and about 10 blocks from St. Joseph's Church and School, which is where we spent a lot of free time. The church was built in 1965 and the pastor at the time, Father Charles Young, asked Dad if he would be willing to clean the church in preparation for the first Masses to be celebrated there, or until they found a janitor to keep it clean. It must have taken the church 18 years to find a janitor, because that's how long Dad and the boys, and sometimes the girls, cleaned the church. Throughout that time, Dad's primary job was as a truck driver and heavy equipment operator at the Marine Base. He held this job for over 28 years, never making more than $13 per hour.

Every Saturday morning, when the boys were helping Dad clean the church, the girls were at home cleaning and rearranging the furniture. This was a ritual that, to this day, we never quite understood. Why did Mom always want to move the furniture on such a frequent basis? Maybe it was her way to ensure that the girls were doing a good job cleaning under the couches and chairs!

Although we had little in the way of money, as children we truly never realized it. We had a very happy home life. Not to say there were never problems—there were. With eight siblings there were always dramas unfolding regularly. But as a whole, we really enjoyed the time together. It is amazing how many new games we created with the pieces of old games or with odds and ends that we found around the house. From a painting easel, piece of cardboard, coat hanger and yarn we were able to make our first indoor "Nerf" basketball hoop. We played with baseballs long after the cover had come off. With eight creative minds and not a lot of money, we easily found ways to have fun.

Dinner in our house was always a pretty big production, and not because it was elaborate. Actually it was quite the contrary. When Mom made dinner (which was every night because we didn't have any extra money to go out to eat) she would typically make one core or staple item. If we were having tacos then that's all we had. If it was soupy potatoes, then that was it. The only exception was tortillas, as they accompanied almost every meal. We never had a salad before dinner, or side dishes, or even desserts. It was simple home cooking, just the way we liked it.

We thought our family space issues were resolved once we moved into the Fredricks Street home in December of 1969. Elaine was soon to be married, so Mom and Dad felt as if they had things under control. Finally, our family size was shrinking rather than growing! However, the shrinking that related to kids, not stuff, as we had more "junk" than most people. Since we had no garage, we had a difficult time storing all of our bicycles, tools, and "stuff" that a family of ten will naturally accumulate. Therefore, we had many *cuartitos*, which is the Spanish word for "small storage shed." It became a family joke, because, at one point, we had more *cuartitos* around the house than kids living there!

Neighborhood Friendships

With such a large family we were able to spend many hours together, some forced but mostly by choice. Whether we were completing our everyday chores and duties or just plain having fun, you rarely found an Olguin alone. To make matters worse for our parents, it always seemed that our house was the gathering place for many of the neighborhood kids, who probably figured that they felt as if nobody would notice the extra bodies! In retrospect, I think we were all proud that our friends were welcomed in our house no matter what time of day.

No matter where we lived, we always seemed to have a few families with whom we developed close relationships. On Nancy Street, there were the as Yslas, the Roddys, and the Beverage families. On Fredricks Street there were the Limebaughs, Estradas and Enriquez families. To this day, we can mention any one of these family names at an Olguin gathering and immediately generate a host of memorable stories and events. .

There are so many great memories from Nancy Street. Ruben and Michael used to camp outside in front of the house with the neighborhood boys, which always turned out to be a great adventure. Once Mom and Dad went to bed and the lights went off in the house, the guys would head out to walk on Main Street to Soutar Dodge and Retzlaff Pontiac to play with and around the cars. We never had a thought in our mind that we might be doing something wrong—like trespassing!

Gary, Ruben and Michael spent their days in the desert above Barstow Junior High hunting for lizards or walking Main Street looking for something interesting to do. Like A favorite pastime was walking into all the gas stations and auto parts stores asking for decals of STP, Bardahl or Champion Spark Plugs decals to post in their bedroom. We sure knew how to have fun.

Living in the desert promoted a healthy amount of water-fights with all the kids in the neighborhood joining in. During one water-fight, our neighbor, Liz Yslas, was shaking water out of a glass milk bottle and hit Ruben on the forehead causing a huge cut and sending blood everywhere. He came running home for Mom to fix it, as she could fix anything. Well, in this case unfortunately, even Mom's miraculous powers were limited, to rush him to the emergency room where Dr. Musgrave put six stitches into Ruben's his head. Then it was on to the next water-fight!

Another fond memory included collecting empty Coke bottles throughout the neighborhood (Mom wouldn't buy Coke for us—it was too expensive—and returning them to the Cottonwood Market for candy or money for the public pool. We also remember some unbelievable carpool experiences with the Yslas family. Getting everyone to Mount Saint Joseph's School in the morning often meant cramming us and the Yslas kids into cars that were made for six or eight people. Fourteen kids in a station wagon—without a seat belt in sight—that was quite an adventure!

It seemed that in the evenings, especially in the summer months, we were rarely in the house before dark. We played what seemed like every game in the book, from hide-'n-seek and kick-the-can, to good old baseball and football. We would be outside with our friends from the time we got home from school, with a short break for dinner, until we heard Mom yelling for us to come in for the night.

Memories of New Mexico

Every summer, Dad and Mom would pack up the car and drive all of us back to Mom's home in Belen, New Mexico, for an extended visit. Dad would stay for a couple of days and then turn around and go back to Barstow to work. The rest of us would stay in Belen for about six weeks. This was our annual summer vacation.

Another interesting adventure occurred whenever Mom was pregnant and getting close to delivery time. The intended plan was for Dad to take Mom to her mother's home in Belen about six weeks before birth, so that she could stay with her mother until the baby was born. However, for various reasons, the plan didn't always come to pass, and sometimes Mom would end up staying in Barstow for the arrival of the next Olguin baby. The ironic conclusion was that whenever Mom stayed home in Barstow, she had a boy, and whenever she would travel to Belen, she had a girl!

In Belen, our grandmother lived in an old house next to the railroad tracks where we would stay. The area was very green with many trees and an irrigation ditch running behind the back of the house. Along with the ditch came mosquitoes, and they must have been mosquitoes that didn't like visitors because we seemed to be the only ones getting bit. Maybe those mosquitoes saw us as new California blood!

Elaine Olguin at Vina's Mom's house in Belen, NM, 1954.

There was much to see and explore in this environment. It seemed as though every day was a new adventure. Next door lived Aunt Concha, Mom's older sister, and we spent much of our time there.

We hope you enjoy
Elaine Olguin Villareal

Her house was a huge barn transformed into a home, so there were many rooms and floors to explore. Lori got really scared when she was allowed to spend the night in this eerie house and Mom would have to come pick her up.

Mount Saint Josephs

Our next-door neighbors, the Yslas', convinced Mom and Dad that "good Catholics" should have their children in Catholic school. Despite their challenging financial situation, Mom and Dad worked some kind of deal for five of us kids to go to Mount St. Joseph's Catholic School.

On the first day of school, we took a picture of all of us the kids in their uniforms. Gary was in 7th grade; Lori was in 5th grade; Ruben was in 3rd grade; Michael was in 1st grade; and Natalie was in kindergarten.

We used to attend Mass every morning before school. Mass also included the infamous "uniform check" for girls—when you knelt in the pews, the length of your uniform was expected to touch your calves. Of course, the older girls immediately rolled up their skirts at the waist as soon as they got out of church. After church we would head back to the school grounds to stand around the flagpole and begin singing the Mount St. Joseph School Song. The tune is still fresh in our memories! After nine years of attending Catholic School, the tune is deeply embedded in my mind: "To Mount St. Joseph's…"

Natalie had the most time at Mount St. Joseph's School—nine years—so she has some of the best stories. Natalie was a crier. One of her most vivid memories was in her first year of kindergarten in Catholic School when Mom was trying to drop her off at school, and she was crying and crying. For the first few days, Sister Teresita, was very patient with her crying. Then, however, after the third day, Sister told Natalie that if she did not stop crying she would put her in the closet—and "You can cry all you want while you're there." Needless to say, the tough love trick worked! Natalie never cried again.

Another kindergarten memory for Natalie was when they we were going to take school pictures. Sister Teresita told the class to make sure to wear their hair combed back out of their faces. So Natalie went home and told my mom, "Sister said to wear my hair out of my face and in a ponytail." So Mom very nicely did her hair slicked back in a ponytail. The actual picture made her look "bald" compared to her usual appearance with hair! Yet, we all remember the famous words that we always took very seriously at Mount St. Joseph's: "But Sister said..."

Natalie also remembered in first grade when the big event was to make your First Communion in the second grade. She had a beautiful white dress, white veil, and white shoes. She had recently gotten her hair cut short in a Pixie cut. So Mom decided to style her hair with pin curls using lots of bobby pins. When she combed her hair out, she had an "Afro" for her First Communion! She remembers being horrified looking in the mirror, and John laughing at her. (It wasn't the first or last time that John laughed at her!) She was in tears because of her hair, but Mom insisted that she looked beautiful for the Lord. Later, when we they got her First Communion class photo, for those who received First Communion, John happened to be mad at her and took a pen and blacked out her face in the class photo.

Ready for first day at St. Joseph's school in Barstow, 1966. L. to R. Lori, Ruben, Natalie, Gary and Mike.

The most coveted prize for Natalie from grade school was in 6th grade when she actually beat Michael in an elocution contest. Michael was always used to beating her at everything, and he was a grade ahead of her. The elocution contest involved reciting stories or poems in front of the entire PTA audience. There were probably 200 people in attendance. Father Keith was Natalie's teacher, and he was in charge of the entire contest. He stressed practicing the speeches over and over and over in front of a mirror. Michael didn't feel that he needed to bother with too much practice. Natalie's story was entitled, "Love." Michael's story was called, "Look, God." Natalie delivered her performance without any problems and felt that she had done a really good job. Michael began his presentation in fine form. However, he had a mental lapse and needed help from the prompter to remember his words. Then he started over from the beginning to get a fresh start. Again, he drew a mental blank and couldn't remember his lines. He finally got so frustrated that he ran off the stage crying. Ultimately, Natalie took first prize in the whole contest, and felt on top of the world! After 27 years, she still enjoys reminding Michael of this event!

For most of her grade school years, Natalie was on the drill team and the Mexican folk-dancing team. The team performed all the time at Barstow activities as well as surrounding locations like the state fair at San Bernardino. The leader was Sister Teresa, and she was fun but also business-like when it came to practice. "I got smacked several times on the back of the calves with a paddle for some reason or another," said Natalie. Being a participant on these teams taught some discipline and some leadership skills, as well as providing an opportunity to share some good times with family and friends.

The tradition of going to Mount St. Joseph's continued with the grandchildren, who also have some of the memories of Drill Team, Mexican folk dancing and "Sister says." Mom and Dad continued to attend all their events, just as they did the children before them.

Growing Up Catholic

Mom and Dad had the amazing ability to make every one

of us feel special. It stemmed from their strong faith in God, their love and respect for each other, and their huge commitment to the Catholic Church. A typical day would start off with Mass, would include prayers before meals, and would end with prayers before bedtime. Those were the fundamentals of our family.

It seemed that Catholicism and being an Olguin were one and the same. From the beginning Mom and Dad built their marriage on the teachings of the Church. They actually chose to cut their honeymoon short so that they could finish a novena—nine Fridays of adoration at the local Belen church!

There is a remarkable story that captures the depth of Mom and Dad's Christian faith. As a teenager, Gillie accidentally fell

Our Lady of Belen Catholic Church in Belen, NM, an important part of the life of the Olguin Family. Painted by Gillie Olguin.This is a photograph but Gillie Olguin painted this chuch for his mother, Manuelita.

off the Belen theater marquee where he worked. He hurt three disks in the center of his back. As a result of this fall, Gillie was unable to walk unassisted down the aisle when he and Vina got married. On the last Friday of adoration, my grandparents took Vina and Gillie to a small church in the mountains of New Mexico called "El Santuario," which was believed to have special healing powers for the lame. In the church, Gillie dusted his feet with the clay soil from the earthen floor of the church sanctuary. The following morning he miraculously rose from bed unassisted! From that day forward, Gillie never suffered from any further ailments of his legs.

Mom and Dad's sense of faith was deeply woven into our daily family routine. As young children, we would not leave for school in the morning until Mom gave us a blessing. I remember as a young child not been able to leave for school until my mother would give me her blessing. We would stand at the door asking in earnest "*la bendicion mama*," refusing to walk out the door until we were blessed. Mom would be involved with washing dishes or laundry or some other housekeeping menial task, and she would have to stop what she was doing to bless us and send us on our way to school.

Another vivid instance of our parents' steadfast belief in God occurred in January 1978. Dad fell victim to a stroke. In the early morning hours, Dad was rushed to the hospital and was near death. Some of us heard our mother praying in the hospital chapel: "His will be done." She prayed, "If it is your will that Gillie should return to you then take him Lord." Elaine, having heard Mom's prayer, responded in frustration, "Mom, what are you saying, are you crazy?" At the time in 1978, we still had five children at home and Dad was the sole source of income for the family. Mom's deep faith in God's will always prevailed in any situation. Dad recovered and, 23 years later, continues to live a happy, healthy life.

We knew as children that we were always under God's care. We could never keep anything from Mom if we were ever in trouble, Mom knew about it the minute we walked into the front door. She had a keen sense of understanding and insight along with her strong Catholic faith. Gary once asked her how she knew when one of us had done something wrong. She always told us that a little angel had told her. Some of us would love to have caught that little angel! "I knew that if I ever found that little angel I was going to kill it," said Gary.

As young teenagers growing up in our parents' house, we learned that Mom and Dad placed our personal decision making into God's hands. There were many times when one of us had to make a choice between something we felt we wanted to do, but which our parents knew was not good for us. Mom and Dad placed us in a soul-searching situation by adding this standard statement prior to making the choice: "Let your conscience be your guide." Even as young adults we always knew what that meant. We were going to be held accountable for what we did, even if our parents were not there to know we did it. This simple but effective approach has served us all well. Now as adults we all still profess our faith and adhere to the teachings of the church.

Memories of Mom

Growing up in the Olguin home, there was never a dull moment. Mom was probably most responsible for the atmosphere of fun that surrounded the family. She had a wonderful sense of humor that was regularly displayed, and it wasn't unusual to hear a lot of laughing around the dinner table. The local priests were often guests in the Olguin home and, on more than one occasion, Mom served them a cup of coffee in a mug that had a ceramic frog built into the bottom looking up at them. She loved to see the looks on their faces when they suddenly spied the frog looking up at them.

Vina in 1948 in Belen, NM.

Mom learned to cook growing up in a large family with lots of brothers. What she learned proved to be valuable when she began cooking for her own family—how to make large meals on a very small budget. Her most common dishes were usually one-pot meals like beans, potato soup or spaghetti. New Mexico is famous for their chile peppers. Therefore, chile was a staple in the Olguin diet. Mom and Dad both grew up on it, and their kids were no different. We ate it with everything!

The best part about green chile was the process of roasting the peppers. It was an all-day affair. First, a huge burlap sack of green chile was purchased and then it was roasted either on the grill or in the oven. Once the skin of the chile was charred black, it was peeled and seeded. Then it was packaged in plastic bags and put into the freezer for year-round enjoyment. Mmm, good!

There are a lot of sayings in the Olguin family that come directly from Mom. Probably the most disliked, yet remembered, saying of hers was "Let your conscience be your guide" was always the most thought-provoking. All the kids still remember that one. It's easy to remember and probably not always that easy to follow. Another one of her favorite sayings was "No mon' no fun, so sad your Dad." That was just her playful way of saying: we can't afford it. The most profound words that she frequently expressed were "The Lord will provide." That statement proved to be true for the Olguin family. There was never a lot of money coming in, but somehow, all the kids always got the special birthday present they were seeking, and we all had the everyday things that were absolutely needed.

Mom's faith was the "secret ingredient," probably the one thing that carried her family through the years, and it's still the glue thing that keeps us together today. Mom was very active in the Church. She led prayer meetings, served was on lots of committees, taught religious education and often planned and attended charismatic conferences. She went to Mass every single day. Her devotion was amazing!

Being active in a church is one thing, but Vina put her faith to practice every day. She was extremely giving. On several occasions, she took people into her home who had nowhere else to go. And even though her own family was just getting by on a very limited income, she had no second thoughts about giving away the extra to someone in need. A very powerful lesson was

Vina and Gillie playing around while camping in 1988.

learned by all of Mom's children just from being raised by her and witnessing the love she continually shared not only with her family but also with her fellow man. There is a passage in the Bible that speaks about feeding the hungry, tending the sick and clothing the naked. Mom took this passage to heart and passed this generosity of spirit on to her children. It was a powerful lesson for all of us! We all strive to be more faithful people by following her example.

In 1991 Mom was diagnosed with early stages of Alzheimer's disease. She gradually got worse. Through it all, she continued to smile, laugh and enjoy the things that were most important to her—her kids, her husband and most importantly her faith. She was certain that God had something wonderful in store for her. Her faith made a lasting impression on all of us. When it came to God, we believed every word that came out of Mom's mouth. Mom suffered with this awful disease for nearly 10 years and, on October 17, 1999, she went home to heaven to be with God.

One of the best things about being part a big family is that there is always someone there when you need them. This was especially true while we were tending to Mom through her long ordeal going through this disease. We were all able to lean on and comfort each other when the times became really tough.

Mom was a wonderful wife and mother and with Dad, gave us children love, faith, freedom and wings to soar! As Mom always said when we were leaving the house, "Let your conscience be your guide and you will never be lost."

Our Family Then—Our Family Today—Growing Up Rich in Love and Poor in Wealth

Growing up in Barstow in the 1960s and 70s was very similar for most people. There were very few wealthy families. Most residents came from lower middle class, blue-collar families. It was common among our friends to have their fathers work at one of three places: Santa Fe Railroad, the Marine Base or Fort Irwin Army Base. Our dad, in particular, worked at the Marine Base as a truck driver. He made a modest wage but truly loved his job. However, with eight children, he was always looking for ways to supplement his income and for most of our childhood he also had a couple of part-time jobs too—anything to make *ends meet.* I am not sure if the ends really ever met, but at least he was able to pull them closer together.

What we always had—which was priceless—was a deep and unconditional love from our parents. Mom and Dad were not overly affectionate with us, nor overly lenient with us—they ran right down the middle. They could get on our cases when we needed it and also be there for a hug when that was needed too. They always encouraged our interests (as long as it didn't involve a great deal of money!). They led by example. As a result, think we all grew up with deep respect for our parents and, but even more importantly, we admired them very much.

The positive influence of our parents is evident in all our lives today. Elaine has been married for 31 years and continues to reside in Barstow with her husband, Rudy. They own a local business. Their two grown children live in Barstow. Marissa works at a retail store and has returned to school, and Melanie is a dance instructor and a nail technician. The first Olguin great-grandchild, Marissa's son, Calob Villareal, lives in Barstow.

Gary and his wife Nancy, along with their 16 year-old son,

Shane, are lifelong residents of Barstow. Gary recently completed his Bachelor's degree in Education and has become a teacher at Henderson Elementary School. He is currently pursuing his Master's degree as well.

Lori also resides in Barstow and works as an office manager. She is proud of her two daughters. Janeese is attending college in San Diego and Lauren will be a junior in high school.

Ruben has settled in Porterville, California, with his new bride, Leslie, and their "Brady Bunch" family. Both Ruben and Leslie are involved in the real estate business. Their kids, ranging in age from 23 to 6, include Ian, Garrett, Evan, Celese and Madison.

Michael graduated from San Diego State and has been a long-time resident of the San Diego area. Michael owns a successful public relations firm. He and his wife Theresa are the proud parents of one-year-old Sophia.

Natalie has been married for 13 years and lives with her husband, Dave, in Manhattan Beach. They have three grown children between them; Kristin, who recently graduated from San Diego State; Scott, who recently graduated from Pepperdine; and Jenica, who is attending community college.

John and his wife, Marisa, live in Torrance with their daughter Hannah, age 4, and their son, Zachary, age 2. John is the Director of Publicity for the Los Angeles Dodgers and remains very popular among the Olguin family, especially with his access to free Dodger tickets! John got his Bachelor's degree from Redlands University and his Master's degree from the University of Richmond.

DeAnne, "our baby," has actually outpaced all the siblings in terms of geographical residences. She and her husband Greg have lived in Phoenix, Denver, Chicago, Frankfurt, Germany and now San Diego. DeAnne followed in Michael's footsteps and graduated with her Bachelor's degree from San Diego State. She currently works as a graphic artist and designer in the San Diego area.

Gillie continues to enjoy life in Barstow. As an energetic "youth" of 71 years, he owns several horses and can be seen almost every day working at a local Barstow ranch—caring for and training a full stable of horses. Someone recently accused him of liking his horses more than his own children, and he commented, "The horses don't ever criticize you!" Actually, between visiting his eight kids and their spouses, plus his sixteen grandchildren and one great-grandchild, he stays a very busy man.

Christmas at Dad's house. L. to R.., Ruben, Lori, Elaine, Mike, Gillie (Dad), DeAnne, John, Natalie and Gary.

Through it all, Dad continues to be respected and admired by all members of the family. His love, guidance and influence have provided all eight of us with more riches than money could ever buy. We continue to have a strong bond among all the family members, and we still love to get together just to have fun and remember the "good ol' days." Oh, and one final thought—all of us still refer to Barstow as "home."

There are definitely some negatives in growing up in a large family. At times it was even difficult to see any of the good things. However, as we have all grown older, it has become more and more apparent how awesome it was. Even to this day, I think we would all say that our best friends could be found within the confines of our family unit. I don't think any of us would change our Barstow upbringing for anything in the world.

Above, Gillie with his great-grandson Calob in 1999 in Barstow.. Right, Gillie's Dad, Jose David in Belen, NM in 1906. **Family love always continues.**

Part 3
The Yslas Family of the Mojave Desert

Top, passport picture of the Yslas family in 1920. L. R., top row, Serafina, Louis, Mucio. Middle, Steve, Petra and Pedro. Bottom, Baby John, Julia and Francisca "Chica". Above picture, John and Paula Yslas, just married. Right, John and daughter Katy Yslas in 1945. Three generations of the Yslas family are pictured here.

#1 Julia Yslas Romero

The Yslas Family

Memories

by Katy Yslas-Yent, Daughter

Growing up in the home of John and Paula Yslas was like a fairy tale. My name is Katy Yslas-Yent and I'm their eldest daughter. There are so many things that I remember about my childhood. So many things that I just took for granted-and I thought every home had what we had. I thought every mother and father went to church every morning. I thought every mother and father kissed each other in the morning and at night before they went to bed. I thought every mother and dad danced in the kitchen to Mexican music. I can tell you that I was extremely lucky.

We lived at 1470 Nancy Street-right next to the park. Later on I'll tell you about how we got this home and my folks still live there.

Mom and Dad met in Seligman, Arizona, where my dad arrived there to work at Santa Fe. He saw my mother and instantly fell in love. My mother and father have been married for 60 years and that tells you how old I am. I was born exactly 10 months after they were married. I have five brothers and one sister. I have brother John Thomas Yslas, Jr. He is two years younger than I am. I remember when John and I were young and my dad used to come home from work and he would open up the door and yell at us, "Go get the pillow, kids!" When he said that, that meant we were going to the drive-in. That was a wonderful, wonderful time for us. I think that's why I like movies now. We would always go to the movies. That was our form of entertainment. In fact, when we were young, my mother and father would send us to the movies every Sunday afternoon-it was the only time they had to be alone. I think that's why they had so many kids! My father worked for the Santa Fe railroad in Barstow for 30 years, raising and supporting a large family.

It was a blessed family with seven children, five brothers and two sisters. It is funny about my brother's names: John, Peter, Paul, Andrew. And between John and Peter came Fred-

Frederick Joseph Yslas. When people asked my mom why she didn't name Fred after an Apostle, she said she didn't realize she was going to have so many boys. Anyway, Fred was the next boy. Then came Peter. By that time I already had two brothers. I wanted a sister so bad...so when my mother was pregnant with Pete, my dad took her to the hospital and I thought for sure they were going to come home with a baby girl for me. When they came home with a baby boy, I cried and cried, and I always felt guilty. I think when Pete was little, I used to treat him special because it made me feel bad since I cried when he was born. Then came Paul. By the time Paul was born, I didn't even think my mom could have any girls; therefore, I was satisfied with another brother. There I was, the eldest GIRL with four brothers. It's very interesting because those years were so happy, although I did a lot of babysitting.

A lot of things I can't remember but when I used to come home, it was just a wonderful feeling. I don't ever remember my mother and father arguing and most of the time it's because my dad says, that way you don't argue; you just say "*Yes Dear,*", but that doesn't paint a correct picture of my father. My father is a strong-willed, outspoken man. My father and mother love each other very much. His motto is "anything to keep peace." Well, the peace of the Lord is what we had in our home and it was wonderful.

Then when I was a junior at Barstow High School, my mother got pregnant again. I just knew for sure that this time it was going to be a girl and guess what? It was. Elizabeth Ann Yslas was born when I was 16 years. And, poor little thing, I wanted to make sure that she was the perfect little girl. I wanted the curls in her hair, I wanted the barrettes in her hair, I wanted her to wear dresses all the time. And Lizzie wanted to go out and play with the boys all the time. You couldn't blame her since she had four older brothers. There are some things that I do remember that were comical when I think about it now.

When I was 17 years old, I got married. At that young age, I was a mere child myself. I married my childhood sweetheart, and before I knew it, I was the mother of Vernon Eugene Williams Jr., also known as "Flip." The reason I am mentioning this now is that after he was born, when he was probably a couple of months old, my mother showed me some pregnancy vitamins-she was pregnant again. So then, along came my youngest brother, Andrew James Yslas. So that was our family: Katherine Marie, John Thomas, Frederick Joseph, Peter Ray, Paul Andrew, Elizabeth Ann, and Andrew James.

Mom and dad never taught any of us Spanish; the reason was that they would be able to talk to each other without any of us knowing what they were saying!

I remember when we were younger, there was always a special devotion to Our Lady of Guadalupe in our home: we said a lot of Rosaries, lit a lot of candles. And of course, our home was a typical Mexican home with the bright walls. I remember when we had an orange kitchen and a yellow living room. One time my bedroom was red. It was the greatest time and happiest childhood that anyone could ask for. When I was in high school, we certainly weren't rich, but we used to have Friday night dances and my dad would buy me a new dress every Friday. I think I was a little spoiled. I remember when I used to be terrified of the wind, and the reason was, my mother was also afraid, and, of course, she made all of us afraid. When it would get really windy, she made us all go in the closet and pray the Rosary. It's something that I have finally gotten over. When I first married my present husband, I was still afraid of the wind and I would hear the wind and I got under the covers; it's something that he broke me of.

Things happen when you're younger, and they just carry on with you. Like my mother constantly helped others, and bless my father's heart, he went along with the program. My family seemed to have a caring concern for others.

My father was the backbone of the family. He was always there when we needed him, always there for the family, always there for my mother, always there for me, always there for my brothers and for my sister, and for the next two generations. My father has been a strong-willed man.

They first moved to Barstow in 1943; then my father was drafted, but returned to Barstow in 1946. They first lived in a housing complex called the Victory Homes-also known as the

Cracker Boxes. In about 1950, they wanted to buy a home but at that time there was still discrimination and prejudice in Barstow. When my father and three other Mexican families tried to buy homes at what was called the Folgesong homes, they were denied. They had approved loans but were still denied. My father did not take this lying down-he fought! He went to the Veteran's Administration and finally got a court order and he was able to purchase their home.

My mother has always been a saint, kind, soft and loving and wonderful to be with. And how did a union like this become a reality?

Let's start with my mother. My mother's name is Paula Petra Rincon. She was born June 29, 1920, in Seligman, Arizona, to the parents of Celestina Ramirez and Juan Rincon. This was during the post World War I depression; it was very hard for the Rincon family. A working family could not make a lot of money then. They did not have much money, so the children never had many toys to play with. Being resourceful though, they made do with what they had-toys such as jacks and sticks from the trees for baseball. One fond memory each speaks about is that the children had no worries to think about; they

Celestina Ramírez Rincón, mother of Paula Yslas.

Aunt Toni and Paula Rincon, Arizona, 1922.

grew up having a happy childhood, which, in turn, sent that message out to their own offspring.

Celestina Ramírez Rincón, Paula's mother, was a hard working housewife who cared for her eight children. The children each had a daily task to perform for her household. One would do the cleaning, one would do the cooking, one would do the ironing-eight of them had their own chore to do. Mercedes (Mozelle) remembers that they had to boil the water on the stove because it was depression time and good water was scarce. Afterwards, they put it in the "baby" tub to bathe in. Each one of the eight bathed in the same water. Celestina also did the laundry

Manuel Ramirez, Paula's uncle, in Seligman, AZ, in 1927.

for the motel owned by Fred Harvey. The military often stayed at this motel, and Celestina did the laundry and ironing for soldiers while still taking care of her own household. She was a hard working individual. The children always dreaded when she put her white scarf on top of her head tied in front of her forehead. The kids knew that she meant business when the white scarf was worn. She was all business when making their own laundry detergent, bleach, and starch. Celestina, a devout Catholic, belonged to the choir and many other organizations.

Juan Rincon, the husband of Celestina and the father of these eight children, was a skilled carpenter who helped struc-

Young Juan Rincon in Mexico in the early years.

ture the communities near the Grand Canyon in the early 1900s. He would come home when he could, and always on the weekends to spend time with his family.

The children of Juan and Celestina were Antonia, Paula, Albert, Mozelle, Josie, Lily (Judge), Caroline, and Manual and Mary who were twins. Manual had an individualistic personality.

Young John Yslas (No. 1) and sister Julia and family members in Clovis, NM about 1937. Later in Seligman, AZ., he worked as a cook at the Harvey House making $4 a day and paid $21 a month for room and board. Here he met Paula Rincon and saved "the best pieces of leg-of-lamb," for her.

He had to have his own certain plate, cup, spoon, fork and knife for each meal. Albert was Celestina's favorite child. The older the children became, the more responsibilities they would have to take on. Due to this, the older siblings had some resentment for the younger ones. But, all in all, they got along as brothers and sisters with a good sense of humor. The prankster of the children was Lily, who was always playing practical jokes. For example, he would put rocks in with the baked goods on the table, and would try to eat them, complaining that they were hard as rock! Albert, also mischievous, would pinch the girls.

Although the family was large and times were hard, they all learned the meaning of love and the importance of family.

My father was actually raised by his brother, Louis. His mother passed away when he was eight months old, and his father passed away when he was still young. My father was the youngest of seven children, born in Mexico on June 24, 1920. His parents were both born on farms that their parents owned in Yahualica, a small town in the state of Jalisco, Mexico.

Our grandfather, Pedro, the father of John, was born around 1875 and died in 1930 in Clovis, New Mexico. Our grandmother, Petra, was born around 1883 and died in 1921 in Clovis, New Mexico.

Fraternal great-grandparents were Valentine Yslas and Paula Villegas. Fraternal great grandparents were Jesus Yslas and Pomposa Elizalde.

Pedro and Petra Yslas came to the USA for the first time with their children in 1908 or 1909, approximately. Pedro was employed by the Santa Fe Railroad as a utility worker, janitor, etc. at the passenger depot in Clovis, New Mexico.

The Santa Fe Railroad had just built this part of their railroad line, so the town of Clovis had started while New Mexico was still a territory. Our grandparents remained in Clovis until 1912, then they returned to Jalisco to visit their parents. At that time, the Mexican revolution was in full swing. They returned to Clovis in April of 1913.

After they returned to Clovis, they stayed until 1918, and then returned to Guadalajara, Jalisco. Alejandro, the oldest child, had graduated from Clovis High School by that time. He got a job with a silver mine company located approximately 80-90 miles from Guadalajara. The mining company also employed our grandfather to run the company store.

Our grandparents again decided to return to the USA. By that time, a passport was needed in order to come into the US.

When our grandfather obtained passport, pictures, etc., they crossed into the USA on September 30, 1920, and have been here ever since.

Five Brothers Growing Up

by John, Fred, Peter, Paul, and Andrew Yslas, Sons

Each of us five boys (John, Fred, Peter, Paul, and Andrew) growing up with two sisters (Elizabeth and Katy), were taught by our parents, John and Paula Yslas, to respect our big sister and take care of our baby sister. As far as our relationship among each other, the boys had a basic rule: all things being equal, the older brother makes the decisions. This helped keep the peace, and kept what could have been many fights, to a minimum.

Paper route: Starting with the oldest brother, John, Grandpa (John Sr.) assigned the job of having a paper route. This job meant preparing papers and "throwing" them every morning by 5:00 A.M. The paper route job was passed on down through the brothers. It should be mentioned, however, more often than not, that Grandpa was the one who did the job! This was his "3rd job."

His first job was at Santa Fe, his 2nd job was janitor at St. Joseph's School, his 3rd was "our" paper route, and his 4th was working at a Richfield gas station on the weekends.

Over the years, he always held at least two of the above jobs. However, his sons could always count on hearing him in the stands at all of our sports events, starting with little league and through our high school years and college-he was always there -you could hear him in the stands.

The most important lesson to all of us has been that, through all of our ups and downs, no matter what the circumstances, we could always count on our parents being behind us 100% with unconditional love.

Something that happened just today-they were visiting with their latest grandchild and he wanted to know why White Sox, their family dog, was not as frisky as usual. His father told him, "White Sox has puppies in her tummy." After a slight pause, he replied, "Did she eat them?" I thought about all the times when Grandma and Grandpa heard similar expressions from their many grandchildren.

Amazing

by Peter Yslas, Son

God has blessed me. I consider myself the luckiest guy in the world, growing up as John and Paula Yslas' son, and I know my brothers and sisters feel the same way. Have you heard of the expression "poor little rich child?" Well, we are known as the luckiest poor children. They give each and every one of us unconditional love. A divine balance that can only come from our Lord.

A few of the many great memories etched in my mind:

When I was a young boy of 12, I desperately wanted an electric guitar. A friend of mine named Nickey Garcia owned a very expensive Fender guitar with an amplifier and microphone. Nickey put together a band with a lead guitar, bass, and a drummer. They needed a rhythm guitar to complete the group. I wanted that vacant position. I knew my dad would never approve of this type of investment, simply because it was far too expensive. So I asked my mom. When it comes to money, my mom is the only person who can change my dad's mind. After a lengthy discussion with my dad, my mom approached me with the good news. Dad said, "Okay! If we could find a way to do it." My mom is the type who always sees the glass of water *half-full*. She assured me she would find a way, and she did. The way she did is, still to this day, amazing to me.

A traveling salesman came through Barstow once a month. He drove a covered truck that looked a little like a milk truck; only it was painted brown, with huge florescent letters reading "Jewel." We called him the "jewel man". He carried various items such as household goods, industrial equipment, foodstuff, and musical instruments. My mom found a way to purchase a $70.00 electric guitar for $5.00 down and $2.00 a month. It took three years, but they paid in full. Amazing! My parents have sacrificed over and over again for their children. We are truly lucky.

Sacrifices

by Elizabeth Yslas, Daughter

I remember when I was in 7th or 8th grade in 1972. It seemed like everyone I knew-all my friends-had a 10-speed bike.

Naturally, I wanted one-badly. My dad worked three jobs just to keep food on the table for our family. I will never forget how he managed to get a 10-speed bike for me. I also remember (around this same time-frame) my dad allowing me to go to JC Penny's every week to buy a new blouse. My friends always had quite a lot more than that and I remember being envious. Today, however, reflecting back, I realize what a sacrifice it was for my dad to give me that money each week so I would be able to buy something new.

When I think of growing up in our house, with my mom and dad, John and Paula, I just knew that it didn't matter what we did wrong or what we did right, they always loved us and, if needed, forgave us with their unconditional love. They have not changed. They are still the most forgiving and loving people I know.

Memories

by Cindi Yslas, Daughter, Sister-in-Law

Here are some memories about John and Paula I hold deep in my heart. I remember when my husband and I first had our daughter. Grandma (Paula) would call and tell us "bring the baby." We were nervous, new parents and always worried when we would leave our little girl, Marysa. We would take a huge supply of everything and give instructions on this and that about the baby. Finally, one time, Grandma said, "It's OK. We know, we had seven kids."

John and Paula have always made me feel like their daughter, rather than a daughter-in-law. I remember an occasion when I had to purchase a plumbing part from a local hardware store and Grandpa had to give me a ride. I went in the store and came out with what I thought was the right part. First he asked me how much I paid for it. When I told him, he shook his head, went in himself and got the same thing for way less! I loved the way he reacted; I felt it was how he would have been with one of his own daughters.

Prayer: You will rarely see Paula without a Rosary in her hand. As everyone knows, John and Paula enjoy traveling to Las Vegas for relaxation. They pray the Rosary on the way there and on the way back home. One time, however, I remember Grandma telling me that Grandpa was so tired on the ride home she was worried he would fall asleep. She said she said something inten-

Family picnic at the Grand Canyon in 1947. L. to R. back row, John, Paula, Celestina, Josie and Juan. Front row, John, Jr. and Katy.

tionally that she knew would make him "mad." Grandma said he did get upset and because of that, he stayed awake for a safe trip home. She told him after they arrived home what she had done!

Love of neighbors: One Christmas, Grandma had met a young boy at Mass. She felt sorry for him, as he had no family in town, and invited him to spend Christmas Eve with the family. So he would not be left out, Paula took some of the gifts (new socks and underwear) that were to be for her sons, and put them out for the stranger.

Prayer: I have never known anyone with more faith in God than John and Paula. I am still in awe over their devotion to God and the Virgin Mary. It is because of this, I am sure, that they have the love they do and the love they share. I remember my daughter and I sharing how if we could be like anyone, it would be to be like Grandma and Grandpa.

Blessed

by Vernon (Flip) Williams, Grandson

Looking back to my younger years, they were very happy times. Being the oldest of John and Paula Yslas' grandchildren, I feel that I was blessed to have the opportunity to be able to be with them on trips such as taking them to Medugorje, Yugoslavia, Seligman, AZ and New Mexico. There are so many happy occasions I can't name them all. I remember, of course, the holidays-Christmases, Thanksgivings-the birthdays, all meaningful holidays to me. Nothing can compare to the love that they taught us. I feel very fortunate to be brought up by them, my second home. As I was close in age with my uncles and aunts, I felt more like a brother to them than a nephew. I can remember being at the home of Grandma and Grandpa at 1470 Nancy, the same home they have now. I remember them bathing me in the kitchen

The Yslas home in Barstow. This is the post WW II house at 1470 Nancy that the bank refused to sell to the Yslas family because of prejudice as the Fogalsong Division was to be "all white." The Yslas and Arredondo families stood up for their rights as Americans and war veterans and integrated this part of Barstow. For almost 50 years they have been good neighbors and raised their family here.

sink-the same kitchen sink they have now. My grandpa would pick us up from school and after school we went to Rodriguez Market to buy meat for that evening's dinner. Grandpa would go to Rodriguez Market every day. After arriving home, Grandma and Grandpa started to prepare dinner. They would put the Mexican music on in the kitchen, and as the food cooked they would dance together. Their main focus in life was to have faith, hope, love and charity. Thinking about charity, my grandmother has been involved in many organizations throughout the years. I guess you can say that they're the main reason why our family is the way it is today.

It is so amazing to look back and remember people that have come into our lives briefly and see them 20 years later. These people remember going to John and Paula's home and always having a wonderful time. I can just hear Grandma saying, "Come and eat"-and Grandpa saying, "Who's this?"

For many years I have been an Eucharistic Minister at the Spanish mass on Sundays, so I had the opportunity to see Grandma and Grandpa dressed up and extremely happy; that's the mass they always attended because they were also Eucharistic Ministers.

Friendship: My grandparents valued friendship the most-not money or rich things, but rather the friends that they met at church, Santa Fe, prayer groups, and in every charitable organization in town. One of Paula's dearest friends that she has known for a lifetime, states that she has never known Paula to say anything bad about another person-ever!

Family Reunions: Family reunions were exciting times for me. To see my grandmother join my great aunts and great uncles and see them laugh and have a good time was wonderful-and to watch each family grow. When we were younger, we would all pile into a station wagon. I was always in the back. We would travel at night-due to the fact that all the kids would be asleep.

I always knew when everything was okay-when Grandpa would give a wink with his right eye and say "sounds like a winner"-this would mean that everything was okay! I knew everything was fine with Grandma when she would bless us with Holy Water as we were ready to depart for a trip; this meant that everything would be okay.

No one could ever ask for a better Grandmother and Grandfather, and I love them very much.

Unconditional Love

by Vincent Williams, Grandson

When I think of John and Paula Yslas, the first thing that I think of is the unconditional love that they give to their family.

I always loved going to Grandma and Grandpa's house when we were little. I always knew that we would be greeted with love, kindness and food.

Even though they had five children living at home at the time, they made me and my brothers feel special-even though we were three snot-nosed little kids. They made us feel as if we were welcomed even though we made their lives a little more complicated and difficult.

I will always remember waiting for Grandpa to pick us up from school. My brothers and I had a choice of walking home or waiting with my Uncle Andrew for 30 minutes for Grandpa to get off work and pick us up and take us to his house. We would usually pick going to Grandma and Grandpa's house-especially on Wednesdays, because that was "taco night" at Del Taco. Grandpa would buy tacos for all of us.

Sunday dinners will always have a special place in my heart. No matter how busy our schedule was we always managed to make time to go eat Grandma's great food. I especially remember her enchiladas and her green chile.

Birthdays were also very priceless because you always knew that Grandma Yslas would get you a card with a little note telling you how special and important that you are to her and Grandpa. These little notes will never be forgotten.

I am proud that my children, Vinney and Kyle, have high morals and are two fine young men. This is not only a reflection on my wife, Joann, and me, but, it all traces back to Grandma and Grandpa. They did a great job in raising my mother, Katy. They instilled great moral values in her that she was able to pass onto me and eventually to my children. Their legacy will go on forever.

Grandma and Grandpa worked hard to put all of their children through Mount Saint Joseph's Catholic School. Even now, they are 80 plus years old; they still work at Saint Joseph's to pay for my son, Kyle's, tuition.

John and Paula's house is the one place that you can go and feel loved. It didn't matter who you were or what you had done; you were always welcome. Grandma and Grandpa always treated you with love and respect, unconditionally.

Reflections

by Steven Williams, Grandson

Grandpa and Grandma (John and Paula) have made many sacrifices that I hold dear to my heart. One that I especially remember is helping my mom raise my brothers and me. Another major sacrifice was volunteer work that they both did at Mt. Saint Joseph's for my education at St. Joseph's School.

Another "special" memory I have of my childhood with Grandpa is when Grandpa used to pay me a penny for every white hair that I could find on his head. Thinking about it now, I should pay him back for causing most of those white hairs! And, thank you, Grandma, for holding the secrets that only you and I share. From the "sugar baby bottles" to the "Principal's office." Not everyone will understand what I mean by this, but my Grandma and Grandpa will.

Most important, my reflection we share is the love for Jesus, the conversation of his miracles in my life today, his grace and mercy and direction of tomorrow.

I love you, from Steve.

A Granddaughter's Memories

by Jeanette Yslas

Paula and John Yslas are my grandparents and I had the opportunity to live with them. I have some wonderful stories to share. I believe that my grandparents have the best relationship of anyone I know. My grandpa gets up every morning by 5:00 a.m. and he will get ready, make coffee, prepare my grandma's cereal or oatmeal and then wake her up. My grandpa makes the best coffee and I enjoy going over to have a cup when I visit. I have asked him what brand he buys but for some reason my coffee never tastes the same as his.

I liked living with my grandparents because I was able to travel with them to family functions or I was at their house when our relatives came to visit. I had the chance to meet my great aunts, great uncles and second cousins. I can remember one time when I was about twelve I went with my grandparents to San Ysidro, California for my great uncle Alejandro's ninetieth birthday. My grandpa drove and San Ysidro is close to the Mexico border, we missed our exit and ended up in Tijuana, so we had to turn around. We finally found my great-uncle's house and had a wonderful time.

I love my grandma's cooking and I like to cook myself. I have asked her for recipes and she has made these dishes so many times that she does not need to measure ingredients like I do. I watched her make tortilias and she told me to put a hand full of flour, and a pinch of salt, and some baking powder...needless to say I wasn't able to follow these directions very well.

Mis Abuelos

by Marysa Yslas, Granddaughter

Es muy difícil elegir sólo a una persona que sea la más importante en mi familia. Es así porque es imposible hablar de mi abuela sin mi abuelo. Mis abuelos paternos, los Yslas, han estado casados sesenta años este año. Mi abuela nunca ha tenido un permiso de manejar y mi abuelo la lleva a cada lugar que ella va. Ellos tienen siete hijos, trece nietos, y siete bisnietos.

Cada día, ellos van a la iglesia para abrirla a las siete de la mañana. Mi abuela reza mas que ninguna persona que yo conozca. Yo amo a mis abuelos porque muchas veces, si voy a su casa, los encuentra durmiendo, mi abuela con un rosario en su mano. Lo que me gusta más de mis abuelos es que después de todo- todos los nietos y todos los años- todavía están enamorados.

Cuando yo tenía dos años, mi mama trabajaba en Contel y empezaba cada día a las siete de la mañana. Mi papá trabajaba en Santa Fe durante la noche y no acababa con el trabajo hasta que las siete y media de la mañana. Mi mama me llevaba a la casa de mis abuelos cada mañana y mi abuela me daba de comer y mi abuelo me llevaba al parque—cada día. Entonces, cuando yo tenía cinco años hasta que tenía ocho años mi mamá me llevaba a la casa de mis abuelos y mi abuela me daba cereal con más azúcar de la que yo queria, y mi abuelo me llevaba a la escuela. Además, cuando estaba en jardín de infantes, un niño me trató mal y yo temía ir a la escuela. Mi abuela entonces empezó ir a mi escuela para ayudar con los niños durante el almuerzo. Pues, yo fui a la misma escuela en el jardín de infantes hasta el octavo grado y mi abuela iba también durante el almuerzo. No fue hasta que yo era mayor que me di cuenta de lo mucho que mis abuelos han hecho por mí y por toda mi familia. Mis abuelos tienen ochenta años y todavía cocinan casi todos los días para mi familia. Todo lo que hacen, lo hacen con amor. Por supuesto en mi familia, yo creo que mis abuelos Juan y Paula son las personas más importantes y estoy segura que toda mi familia está de acuerdo conmigo. Ellos son ejemplos para mí en cada manera de su vida. Yo espero que yo pueda parecerme a ellos aunque sea un poquito.

Great Grandson's Memories

by Vinney Williams, Great Grandson

I always remember that every Sunday the family would get together at my great-grandmother's house to eat Sunday dinner. I always liked this because no one makes better tacos than grandma and papa. Papa would cook up the shells and grandma would get the rest ready. They would make 30-40 tacos every week, and bring them all in on a big plate and put it in the middle of the table. When I was very little I was a picky eater and did not like lettuce in my taco. Grandma would always put two tacos at the end of the plate with no lettuce. Every Sunday two tacos of the big plate were set-aside just for me, and grandma would always remind me that they were just for me.

Even as I got older and really didn't mind what I eat (kind of shows in my bigger size waist line, but I don't mind) Grandma still puts two tacos aside for me, without lettuce.

There are so many more memories, stories, and reflections that could be told about John and Paula Yslas. The one thought that always comes to mind is their unconditional love. They taught their children love of god, love of family, love of neighbor, and love of country.

Birthdays

by Vernon "Flip" Williams

Birthdays are special to me as far back as I can remember. Birthdays have been happy times for the Yslas family. My mother, grandfather, grandmother and great grandmother and great grandfather have enhanced the unity of the family through the generations by their celebrations of birthdays. They called the birthday children on their birthdays in the morning before they go to work or to school, singing happy birthday over the phone to brighten their day.

When I was in grammar school, I remember cup cakes brought to my classroom for my birthday. We also celebrated birthdays at Grandma's house, because we always ate Sunday dinners there.

It was not until I was a little older that "Mananitas" was introduced to me. That has become part of our family's tradition too.

Mañanitas Tapatías

Qué linda está la manana
En que vengo a saludarte.
Venimos todos con gusto
Y placer a felicitarte

El día en que tú naciste
Nacieron todas las flores
Y en la pila del bautismo
Cantaron los ruisenòres

Yo viene amaneciendo
Ya la luz del dia nos dió
Levántate, virgencita
Mira que ya amaneció

De las estrellas del cielo
Tengo que bajarte dos:
Una para saludarte
Y otro para decirte adiós

Ya viene alborando el día,
Que linda está la mañana,
Saludemos a María,
Bueno dias, Guadalupana.

Quisiera ser un San Juan,
Quisiera ser un San pedro,
Y venirte a saludar
Con la música del cielo.

El cerro del Tepeyac
Escogiste por morada;
Por eso te saludamos,
Buenos diás, Guadalupana.

[In English]
How beautiful is the morning
In which I come to greet you.
We all come with pleasure
And joy to make you happy.

The day that you were born
All of the flowers were born
And at the baptism font
The nightingales sang.

Already the light of the day
God gave us
Of the stars of heaven
I have taken down two for you:
One to greet you.

And another to tell you goodbye.

Now the dawn is coming
How beautiful is the morning
Let us greet María,
Good day, Guadalupana.

Never too old to celebrate Happy Birthdays, as John and Paula Yslas celebrate their 80th birthday.

I would like to be Saint John;
I would like to be Saint Peter
And come to greet you
With the music from heaven.

The hills of Tepeyac
You chose for a dwelling
For that reason we greet you,
Good day, Guadalupana.

The Baby Of The Family

Andrew James Yslas-Son

Growing up as the baby of the family gave me a opportunity to be spoiled, which was a big plus for me. In fact until this day I admit I'm still spoiled, not only by my mom and dad but also by my brothers and sisters.

I was born after my older sister Katy was married and already had a son.

She and her first husband took me everywhere with their

children (my nephews). We would go on campouts, boating, fishing, hunting, all the fun things that kids like to do.

When I came home, my brothers, John, Fred, Peter and Paul and my sister Liz would take me various places. Perhaps it was a competition because I was the baby brother.

My mom, as spiritual as she is, taught me about being spiritual, kind, loving and caring as she still is today.

Celestina and Juan Rincon in front of their home in Seligman, AZ.

My dad kept me out of trouble and kept me in sports. He's a man's man and I respect him highly.

I am lucky to be the baby of the family and grateful to be the son of wonderful parents.

A Wonderful Life

Paula Rincon Yslas - Mother

I was the second child of nine children born to Juan and Celestina Rincon.

I was born on June 29, 1920 in Seligman, Arizona.

I attended Seligman Grade School and Seligman High School and graduated from Seligman High School in 1939. I was an average student and liked school very much. I never got in trouble except once when I played hooky with my friend and I remember smoking a cigarette. At that time it was considered a very bad thing to do, but it made me so sick that I never touched a cigarette again in my life.

After graduation I started working for a couple by the name of Mr. and Mrs.Zeluff. I did all the housework, laundry and cooking for $9 per month. When Mr. and Mrs. Zeluff moved to Los Angeles, I moved with them. I stayed two months and then my mother wanted me home, so I went back to Seligman.

I then started working for Fred Harvey. During that time John Yslas moved to Seligman from Clovis, New Mexico. He worked for Santa Fe. We met, fell in love and then we were married a year later in December of 1941.

Katy was our first born and later we had six more children, and we lived happily ever after.

John and Paula Yslas 50th Wedding Anniversary in December 1991.

John and Paula Yslas wedding photo, December 20, 1941 in Seligman, AZ.

Part 4
The Luna Family of the Mojave Desert

Jose P. Luna built this house in Los Lunas, NM, in 1908. Here, Noberto and Leo Luna were born—three generations of Lunas. Noberto is shown in the lower left at the time of his wedding to Annie Zamora.

Wedding picture of Margarita Gallegos and Jose P. Luna in Los Lunas, NM.

LUNA FAMILY

The Final Sunset
Leo A. Luna

As I gaze out towards the west, I see a fabulous sunset, laden with many colors. It is one of the most beautiful sights, God, in all his wisdom, has created. I also think of the sunsets of our lives. Someday we will have a final sunset, complete with many colors.

Noberto G. Luna, born May 5, 1916 in Los Lunas, NM.

In the normal course of our lives we lose loved ones. Their lives have ended on this earth, they have seen their final sunsets, but their spirits and memories will live forever in the hearts and minds of their families.

This book reflects how each of us looked at our parents, who have seen their final sunset. Hopefully the reader will feel how our beloved parents and loved ones affected, not only our daily lives, but also our future.

The Luna Family

He was born Noberto Florian Luna on May 5, 1916, in Los Lunas, New Mexico, in a house his father, Jose P. Luna, had built of adobe. Later on in life he took his mother's maiden name for his middle name, Gallegos. His mother was Margarita

Annie Zamora shortly before she married Dad.

Gallegos, also from Los Lunas. He was the oldest of three boys. His mother also gave birth to four girls. Dad died on December 19, 1999, at age 83.

Mother was born Anita Feliz Zamora on February 27, 1918, in Tome, New Mexico. Her mother, Josephita Gabaldon, died when Mom was only 11 years old. Her father, Jose S. Zamora, was a sheepherder and was gone most of the time. The children were raised by their uncle, Manual Gabaldon. She was the third of seven girls and one boy. Mom passed away on March 25, 2001, also at age 83, after a long battle with Alzheimers Disease. This is their legacy.

Godfather And Godson

Both of these families were tied together long before Mom and Dad met. As the story goes, when Jose Luna was twelve years old he was asked to be a godfather to a little boy from across the Rio Grande River, in Tome. The details of why he was asked have long been lost. Rather than take the long way around and cross the river on the bridge, he took off his clothes and swam across, where the river was closest to Los Lunas. After the ceremony was concluded all went their separate ways with no idea what happened to the others. After mom and dad met, it was discovered that dad's father, Jose Luna, was the godfather of mom's father, Jose Zamora!

A Slice Of Pie

Dad told us once of an incident that occurred when he was a very young man and my grandfather, whom we called Nino, my dad and my Uncle Miramon, whom we called Uncle Joe, went to Albuquerque on some business. After their business was finished Nino took them to a small cafe where, after eating their dinner, he ordered a slice of lemon meringue pie for himself and each of his boys. My dad and Uncle Joe had never seen anything like that before. Grandma used to make what we call tarts now but was a flat fruit-filled dessert. These two country boys were not up to trying anything new so each one would tell the other, "You eat it," and the other would reply, "No, you eat it." This went on back and forth 'til Nino got mad and told them, in Spanish, "What have I raised here, a couple of idiots? Pendejos! " Nino had to eat all three slices of pie. Later on in life lemon meringue pie was Dad's favorite dessert.

Sinking A Well

Sinking a well in New Mexico in the early 1900s was hard work. Two or more men using sledgehammers literally pounded the pipe down into the ground, adding pipe as required to reach the desired depth. My grandfather, Jose Luna, was a very ingenious man and came up with a way to make it easier and faster. He took an oxygen cylinder, cut the bottom off it and had handles bolted to the sides so two men could lift and pound the water pipes into the ground. Later on he came up with a tripod with a pulley on top to make it easier. A rope went up through the pulley, and then down to the top of the cylinder, which had a ring welded on it. One man would pull on the rope while another man (or two) would guide the cylinder down and pound the pipe into the ground. My dad and his brothers usually did the work of guiding the cylinder down while my grandfather worked the rope. They got so good at this that the cylinder actually lifted over the top of the pipe before being brought down.

As the story goes, one day when Nino, my dad and my Uncle Joe were sinking a well, the boys got to horsing around. Nino stopped and told his sons, "Don't be messing around or you're going to get hurt." That would calm them down for a little while, then they would start laughing and horsing around again. Nino stopped and again told them to stop messing around. Then they continued with the work at hand. Finally the inevitable happened. The cylinder missed the pipe, and came down hard on Dad's knee! He lay on the ground crying in pain, thinking his leg was shattered for life. Nino checked his leg, found it was not broken, helped him stand up and asked if he was all right. Dad tested his leg by putting his weight on it and found he was just sore. Nino asked him again if he was all right and asked him if he wanted to sit on the wagon for a while. When Dad told him he was all right, Nino took him by the arm as if walking him to the wagon and proceeded to kick his butt! Years later Dad and Uncle Joe would sit and laugh about that incident 'til tears rolled down their cheeks!

CC Camp (Civilian Conservation Corps)

Daddy used to tell the story about one night while around the campfire at CC Camp one of his friends had buried a pot of beans to cook. The friend built a really hot fire and placed a covered pot of beans in the fire and covered it with dirt. My dad's friend was known for cooking the best beans underground, and as they passed the bottle around he was doing a bit of bragging. "Nobody cooks beans like me," he boasted.

Well, as the fireside stories started up, one guy was telling a story about a wild bear that had been spotted around the camp, when all of a sudden there was a loud explosion! They all scattered and my dad was on his hands and knees when one of the new enlistees jumped on daddy's back as they rolled down the hill, Dad thinking it was the bear they were just telling about, scrambled for his life...

After everything settled down and everyone regrouped at the campfire, it was discovered that the pot of beans had exploded! It was a sight to see the men all were covered with mud and beans and the men all ribbed the man with "Yeah, no one can cook beans like you." *Told by Noberto Luna to all his children at one time or another.*

How They Met

Saleta Ann Luna

As a little girl I would ask my grandmother how she and my grandfather met. She told me she was working in nursery taking care of children in Los Lunas, New Mexico. One day she noticed a handsome young man walking down the street. Of course, it was my grandfather, Noberto G. Luna. My grandmother noticed that he would walk by the nursery at the same time everyday, so she would rush to the window, peep through the curtains, only to close them quickly as he would go by, so he wouldn't see her. Little did she know that Noberto had his eye on her too.

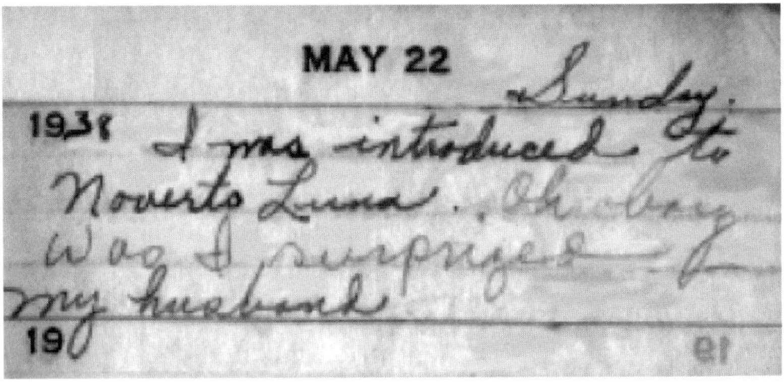

MAY 22

Sunday.

1938 I was introduced to Noverto Luna. Oh boy Was I surprized — my husband

19

Excerpt from Mom's diary. Sunday, May 22, 1938, I was introduced to Noverto Luna. Oh boy, was I surprized. "My husband," she added later on.

Eventually they were introduced by a mutual friend and from then, there was no stopping them. They were married, and stayed married for sixty-one years. They had seven children, one of which passed away when she was an infant (Lydia) whom my oldest daughter is named after. They raised one grandchild and I'm very happy and feel blessed to say was myself. I'm going to close this story by saying life is short but sweet, so every day should be lived as if it were your last and should be enjoyed with the most important thing you could have—your family.

Our Early Years In Barstow

Ben Luna

When we were still in New Mexico, I believe around 1947; a visitor would come out to the house, sometimes on a bicycle, sometimes on horseback. He and Dad talked about work availability in Los Lunas. That is when the talk started about Dad going away to find work.

Shortly thereafter Dad left with Grandpa Zamora (Mom's father) for Barstow. I believe that Grandpa Zamora was working at the Marine Base at the time.

I don't remember boarding the train to come to Barstow, but I do remember looking out the window and seeing the steam engine when the train was rounding a long curve. I remember being very sad leaving my grandmother behind.

I remember arriving at the Barstow station in the early morning and Dad waiting for us, with a borrowed car, to take us to our new home in the projects, which was a trailer in space #42. We later moved to another trailer in space #52, then again to apartment #110, then finally to apartment #109, where we lived for a several years.

Dad got a job right away with the housing project. They told him that they could give him a job for a little while because they were going to tear the projects down in about six months.

Dad was a jack-of-all-trades and he could do just about anything he set his mind to, from plumbing, to electrical, to house building, and even auto mechanics.

The Victory Homes

Leo A. Luna

Work was scarce in New Mexico when Mom and Dad decided to move to California. Times were very tough and Daddy made $7.00 a week. Mama used to sew everyone's clothes and they had to sell her sewing machine, which was very precious to her, to buy milk for Benny. Dad arrived in Barstow in December of 1947, found work, a place to live, then sent for Mom and the rest of his family. We arrived in February of 1948. I was eight years old, Benny was six, Helen was four, Patsy was two, Nobie was one, and Randy wasn't born yet. I remember getting on the train for what we thought was going to be a visit, not knowing we were going to spend most of our lives here. We lived in a housing project, which was called the Victory Homes. A large majority of people who came to Barstow lived in the Victory Homes at one time or another. Dad was working construction but finally got a job with the San Bernardino County Housing Authority working as a maintenance man in the Victory Homes. He came to know everybody who lived there and he made many friends. Mom had a full time job taking care of us kids and our friends whom we brought to our home for any goodies mom might have made. If she hadn't made any goodies we would take a tortilla, spread lard on it, salt it and eat it. Our friends would join us in a "sandwich" before we went out and played some more. We made many friends during our early years and some of these people remain friends to this day.

Shine, Mister?

When my brother, Benny, and I were just a couple of little kids, around eight or nine years old, Dad made us a couple of shoe shine boxes so could go downtown Barstow and shine shoes. He bought us wax, liquid, brushes and shine rags. That way we could make our own spending money. We used to charge a dime to shine a pair of shoes, no matter what color, as long as they were either brown, black, or cordovan. Business was kind of brisk whenever there were a bunch of Marines or soldiers in town. There were other kids that were in this enterprise also and sometimes competition got a little out of hand. I remember once when Benny and I went to town with our shine boxes, one of the other kids had several Marines lined up for a shoeshine. One of the Marines asked my brother to shine his shoes for him, so he did. The other kid wanted Benny to give him the money because "it was his customer." Benny refused and a fight broke out. Since I was his big brother, I came to his rescue and I ended up in the fistfight. The owner of Dillingham's Market came out and broke it up 'cause he thought we might fall through his plate glass window. Benny ended up with the money and I ended up with the bruises.

Being kids we would get thirsty from all that activity, but

we had a good source of drinking water! Swamp coolers (or evaporative coolers as they are appropriately called) were used to cool the apartments but they did not have water pumps like modern coolers do now. A water line went to the top of the

Leo Luna at the Victory Homes, built during WWII in Barstow. Note the coolers in the windows.

cooler, and then water would flow down the pads where air was cooled by evaporation before going into the apartment. These coolers tended to overflow most of the time and cool water ran onto the ground. There was usually quite a bit of grass growing under the coolers due to the water overflow. When we were thirsty we would stand under the cooler and let water run into our mouths to slack our thirst. So most of the kids around there have drunk cooler water at one time or another. I don't know of anybody that died from drinking that water so it must not have been too bad for our health!

Our Childhood

Benny Luna

It must have been in 1949 or maybe 1950 that I played in the little league in Barstow. I remember that the first year we just had tee shirts. The second year we had full uniforms. If I remember correctly, the whole city of Barstow had only 4 teams. The sponsors of the teams were, Santa Fe, Rotary Club, Lions Club, and maybe the Elks Lodge. Most of our games were played in Hobbs field.

In those days we didn't watch television during the day, so we had all kinds of outdoor games, Kick The Can, Marbles, Hide and Seek, Cowboys and Indians, Green Light-Red Light. During the evening we would watch Cousin Herb, a Country and Western singing show, on Channel 10 from Bakersfield.

There was a playground in the project where we would play during the day. Next to the playground was an outdoor theater where they had different sorts of entertainment. I remember Dad hitting fly balls to us to improve our skills in baseball. He also loved boxing and I remember watching many boxing matches. He would match me with different kids at the project for a one round boxing match. Sometimes I would win; sometimes I would get my block knocked off.

Clotheslines

Patsy Luna Biles

Living in the "projects," there is an unsaid pact that is shared by all. The secret is who gets to use it first and that is the way it was with the clotheslines.

Mama had a set day every week to wash clothes, to "sprinkle" the clothes for ironing, and of course "ironing day."

Washdays were always a tough day at home because it was done in the kitchen sinks with a wringer-type Maytag washing machine. That tied up the sink for anything else until the wash was done.

Mama was very particular how her wash was done and how it looked on the clotheslines. I really believe today, your neighbors are judged by how their clothes looked "on the lines."

As children, we would catch parts of conversations and would hear the neighbors talking about someone's clothes "on the lines," good, bad, or indifferent.

Mama would make sure she was up early to beat the crowd for her set wash day to put "Dibs" on the lines she was going to use for the day. Before she would hang her clothes, she would run a cloth on the clothes lines to clean them. She taught my sister, Helen, and me how to hang clothes the "right way." There was a certain way you would hang the towels, and the pants. To this day, when I hang clothes, those lessons come alive again and I can hear Mama telling me how to do it the "right way."

There is nothing that compares to the fresh smell of clean clothes that have been dried outside on the clotheslines.

Today, I have a bag of clothespins hanging in my laundry room to remind me of these wonderful memories.

My Mom And Pops & Route 66

Randy Luna

Vacation time! School is finally out and it's time for play! My dearest memories of Mom and Pops have been the wonderful vacations we took every year during the summer months 700+ miles east on Route 66 to the township of Los Lunas, New Mexico, my mother and father's home in their younger years. It had always been such a slower pace of life in Los Lunas when we finally settled in for the two weeks that we had to visit distant relatives. You could hear the steady drone of the John Deere tractors early in the morning, the absence of traffic noises that allowed you literally go back in time. But to me, the adventure was not always arriving at the destination, but the true adventure of getting there! The times that we left for the "other side of the world" vacation, as it seems were always the highlights of our childhood. Mom and my two sisters packing multiple suitcases, and the consumables for the long trip eastward, Pops with the oldest of the three brothers helping to stash the water bags on the front bumper, airing up the fairly good shape tires, making sure we had enough oil cans to last us to the next town between the long desert towns on Route 66. Packing up the car for the trip was a sight in anybody's imagination today. In this day and age, it isn't even feasible to think about getting eight of us in one car headed down the two-lane road for approximate 700+ miles for 12 hours for a "relaxing" vacation! I do remember the unexplainable, interesting times on the road that we had, such as Mom bringing along most of the contents of the refrigerator to keep us fed during the long trip ahead. We now know that this food was to be eaten many, many miles and hours from California... Reality check, Mom was already breaking out sandwiches for all of us within spitting distance from Newberry Springs, which was about 30 miles from home! No complaints from any of us kids in the back seat! Food was an "Event" for most of us.

Engine noises must have been so relaxing speeding down the two-lane road at all of 60 mph that it would make Pops want to nod off while driving even after a good night sleep before the trip, or so we thought. Think about it, six kids, 700 miles, etc.! Mom would make sure this wouldn't happen by placing wet hand towels on his forehead to keep him alert. I thought the eight of us in the car would be enough to keep even the dearly departed from eternal sleep. I remember on several different occasions feeling the car slowing down, then all of us looking out the window many, many miles from California and seeing blinding red lights of ambulances, police cars and horrific head-on accidents of cars intending to make the trip at a faster pace than usual by trying to pass other slower cars on hills, or curves, with disastrous results. What a sight to see when it was our time to pass the wreckage of an accident. I'm sure we were a sight too, with all our faces glued to the windows filled with horror. At that time I along with my brothers and sisters always seemed to pray to God to see that Pops would see life at a slower pace and not feel brave enough to do what those poor people had just experienced!

Scenery along Route 66 was something that kept us all from boredom. There were no cookie cutter restaurants, fast food eateries, gas stations, and stores along Route 66 as they are now along the Interstate. They were all different from the previous one we would stop at. All were trying to get your business along the old Route 66. But there were many bright billboards and sayings (Burma Shave) that kept you entertained on the long journey. Many of them seemed to work because Pops was sure informed about them by Mom and us when we seen one of interest. The old black jackrabbit stood out as the one I remember to this day. When first seen, he was hundreds and hundreds of miles from where we were. He seemed to get bigger by the mile until my interest to see this humongous giant black jackrabbit was unbearable! I gathered that most travelers on Route 66 had many kids like us because the stores were noisy and crowded. Other fathers didn't seemed interested in the stores like us because they stayed outside and checked over the car while we spent most of our vacation money on trinkets and souvenirs. Maybe that's why fathers here looked sad and tired! The Arizona Meteor Crater was another amazing stop on Route 66 that us kids kept talking Pops in to stopping to see. Such patience this man and woman had with us! We being kids at the time never experienced dangerous, unfriendly times with others that were traveling. Maybe that is the nature of kids during that time. All people were friendly and vacations were vacations.

Many towns we traveled through on our journey were at nights, which turned bitter cold when we took a break from our long road trip. It was such an experience to feel the cold outdoors and see different buildings, people of different nationalities in other towns. A thing of luxury for all of us was the restaurant since we had not ever eaten in them before. Feeding eight people in those days was cheap compared to the present. Mom and Pops would order all the food for us since none of us would dare talk to a waitress back then. Must have been a sight! To think we paid somebody else to cook the food for us. I guess we considered our family as one big happy family and outsiders were strangers.

The automobiles that we took on our vacations were probably in the best of shape compared to others that we saw along the side of the road on Route 66. I don't remember many breakdowns during our summer vacations. Those experiences, being stranded along side of the road we could do without. These were the comforts that we got from Pops in taking care of his cars to carry his family on such trips. Pops was always mechanically inclined and an excellent mechanic. Many of his sons and grandsons followed in his footsteps and continue to help others with their cars.

It seemed that we always arrived at Los Lunas between midnight and 6:00 am in the morning. If it was early morning when we arrived, we all waited outside in the car while Pops walked and knocked on Grandma and Nino's door and fooled Grandma into thinking it was somebody else, then she yelled "NOBERTO! " in tears and happiness when she recognized her oldest son's voice! We had arrived!

The countless funny, scary, touching family memories of our mom and pops and Route 66 come and go, along with the good and the sad times I experience since my mom and pops passed onto Our Lord.

By the time this is in print, two of my brothers, a brother-in-law and I, up on Harley-Davidson motorcycles will have taken a journey from Chicago, Illinois to Santa Monica, California, on old Route 66, on its 75th year anniversary. My oldest brother Leo joined us in Los Lunas. When we enter New Mexico, I am sure that we will feel the presence of our father sitting with us to feel the wind and help guide and share with us the memories of Route 66 through the rest of the trip, remembering our summer vacations times with them, our mom and pops!

Mamma

Helen Raster

"A HELL OF A WOMAN"' is my phrase for showing admiration for a woman who I would like to be or I feel does extraordinary things. I know my mother was my first contact with a Hell Of a woman. I mean this in only the highest of compliments. She is superwoman; she is strong, yet tender, she is fiery yet kind, and best of all, those she loves are loved with an unconditional love. Her spirit is unmatched whether you are on the good side of her or the victim of her unleashed wrath. You are well aware you have met ANITA LUNA.

When I was very young I was, at times, embarrassed by her fiery spirit, as I grew up I came to rely on her spirit, it was something you could definitely count on. Especially if you needed an ally when you felt abandoned by others. There was comfort in knowing your mom would 'kick ass' if anybody messed with us kids.

Mamma would fly off the handle at just about anything. Yet all the kids at the projects used to congregate at our house to sample her fresh made tortillas. She always had time for our friends and us. These are a few of the cherished stories about **Mamma**.

Tide Soap

We used to have really shining hair. So shining in fact, that people, complete strangers would stop us and ask us what kind of shampoo we used. We would always answer honestly "TIDE SOAP." Being five & six years old we didn't know that anybody else didn't use the same thing we did.

One day mom overheard someone asking us what kind of shampoo we used, Patsy and I piped up proudly "TIDE SOAP." She quickly grabbed us by the arm and tells firmly "If anybody asks you what you wash your hair with, you tell them Halo Shampoo."

A few years later when television was introduced to the home, we watched laundry and shampoo commercials and realized that Tide Soap was a laundry detergent not a hair shampoo, but we continued to use Tide Soap.

I can still remember the heat sensation on my wet hair as mamma poured a handful of Tide soap in to her hand and onto my head and scrubbed. At first, the granules would feel gritty against my scalp but the lather would take over and then the rinsing off would burn my eyes as soap got into my eyes. It was then that mamma would rub my clean hair and told me that my hair had to be *'squeaky clean.'*

I used Tide Soap into my teen years and there have been times as an adult and mother I have used Tide Soap when we were out of shampoo. It still has the same results today as it did fifty years ago. I can't see a box of Tide without remembering our shining hair.

Spoons

I remember mealtime was a family affair. As soon as us kids heard "Come and Eat." All six of us kids came a running. Everyone came to eat at the same time and in the same place—the kitchen table. I don't remember a table being set as we know it today, but always in the middle of the table was the brightly colored metal jar of spoons. Along with fresh made tortillas and the main dish, which usually were beans and meat or beans and potatoes. We would all grab our spoon for the meal as well as a piece of tortilla. We mostly ate with a bit of tortilla torn off and made into a little scoop, but there was a few times we would need a spoon to finish off the meal.

One time really sticks out in my mind, as we were eating, Mamma spotted a rather nicely dressed woman walking through our gate, she grabbed the spoons out of our hands and handed us this odd looking thing with sharp points on the end and told us, or rather ordered us, to eat with them.

As it turned out, the nicely dressed woman was a teacher who was making her rounds with her new students. A common practice of teachers just before the school year started. She had come to introduce herself to one of us, and to get to know our parents. As this conversation is going on, I'm still looking at this thing that was handed to me in a hurry and trying to figure out how I could possibly eat with this thing that has slits at the end and the juice would fall through? After the visitor left we were handed back our spoons to finish our meal.

Many years later I found out that most people used a fork to eat with, but still to this day I would much rather have a freshly made tortilla and yes, a spoon from the jar in the middle of the table.

The Rosary Vs. The Radio

We used to have this Philco radio. It was a heavy, sort of brown-blond wooden case that had glass tubes in it. It had a short wave band that received reception as far away as China. At night the reception was stronger. The radio was in Mamma and Daddy's bedroom. We kids used to turn the knob just to hear something foreign. It held our attention for a few minutes and then we were off to play something else.

One evening in particular, Mamma was kneeling in her bedroom saying the rosary. It was probably the only quiet place in a houseful of kids. In comes Daddy to go to bed, and turns on the radio. Mom turns and asks him to turn the radio off since she is praying her rosary. He tells her, "Go in the other room and pray, the radio is in here," she sighs a frustrated sigh and continues praying her rosary passing the beads through her fingers. Then daddy turns the short wave knob, which is very scratchy and loud, and she says once again, "Can you **please** turn off that radio? I'm praying my rosary." He tells her, "I told you if you want to pray go in the other room, I can't move the radio."

She shot him an 'if-looks-could-kill' look and continues passing the beads through her fingers as she prays. Then daddy, who loved to get a rise out of her, turns the radio a little bit louder. Totally frustrated she asked him through tight lips, "I'm asking you again, can you ***please*** turn that radio off I'm trying to pray my rosary?" He says again, "If you want to pray go in the other room, I can't take the radio out of here." Totally exasperated and out of patience she looks up heaven bound and says, "May God give me patience to put up with this damn shit man!"

The Big Flip-off

Told by Mark Gerardi

One day Grandma picked up my friend, Jerry, and I from school. She had to take us to town with her on an errand. We were probably about 12 and 13 years old. We both sat in the back seat of Grandma's '68 Ford. We stopped at a stop sign and Grandma looked one way and started to pull out onto the traffic, well just then a man driving a car lays on the horn, as grandma hits the brakes. "You stupid idiot," he yells at Grandma and she apologized, "I'm sorry I didn't see you coming." I got very nervous because I know Grandma's temper, and the man again yelled, "You should be sorry! You didn't even look my way before you pulled out!"

Grandma again said, "I'm really sorry, I didn't realize I didn't look before I pulled out." The more Grandma apologized the madder the man seemed to get, and I'm sitting in the back seat knowing Grandma isn't going to take his yelling at her much longer.

His final insult was, "You have no business behind the wheel of a car driving like that." When finally it happened, she comes back at him through gritted teeth, "Well to hell with you,

I already apologized once!" and she sticks her hand out the window and flips the man the bird. Jerry and I couldn't hide our shock and busted up laughing as we watched Grandma have the *final say so* and the man's mouth drop open as we whizzed by. Jerry thought my Grandma was the *coolest* Grandma in the world. How many Grandmas do you know that flip people off?

Grandma Luna

Hal Jacob Raster

I think about my Grandmother often. What a rock she was! She made all of us grandchildren feel like we were her favorite but there is no doubt in my mind that I truly was. Grandma was always doing something, if it wasn't cooking; it was making a cake, sewing, playing her guitar—something. It's funny but I don't ever remember Grandma having "idle time." I loved my grandmother's "passion," what an advocate to have in your corner. She took me out job-hunting when I got my first job (dishwasher at Sambo's restaurant). She was so proud. Also when I got my drivers license it was grandma who played a key role in that. She used to save her old DMV driving tests (about four or five different ones) after studying with her I aced the test. I loved the way she called me "Mijo." Talk about "unconditional love," we (those blessed by her love) could do no wrong, in fact I felt sorry for those who might knowingly or unknowingly hurt one of her loved ones, that's when you would see her "passion" unveiled. Grandma was a rock. Strong and tough...no doubt about that... but she was also very loving, and compassionate, her and my grandfather helped many people, not only with money but time, effort, food and most importantly with their friendship. I remember at their 50th wedding anniversary a friend of theirs was making a toast to them and he said, "You determine a rich man not by his bank account but by the number of people who call him friend." In that case my grandparents were abundantly wealthy. I also have another memory of that 50th anniversary I'll always keep in my heart. I was asked to go pick them up from their house (an honor I came to cherish even more as time has gone by) and bring them to the party where their family and friends awaited them. As I stood outside, they came out, Grandma first, dressed to impress as usual...then Grandpa, he was good looking and he knew it...but what I'll never forget is the way that they looked at each other when he came out of that house—what a moment! I'll never forget the way Grandma looked at him, what love! Her face was full of a softness I had never seen. He looked back, so proud, he was happy, his eyes were twinkling, she was happy, she was softly crying, she told him that she loved him and he told her that he loved her, they kissed each other and then they hugged. She pinned his corsage on his lapel looking like an excited young bride. They were so happy and proud, it was such a special moment and they both knew it. They both got in the back seat and off we went to celebrate with their most cherished possessions, their family and friends.

Memories Of Grandma

Stephanie Luna Gallop

Beautiful
Young, and even old
Finely dressed for church
With her turquoise jewelry and painted nails
Playing her guitar
A permanent fixture in the choir

A gigantic house up on a hill
Guarded by the best dog ever
Shep knew every family member
A dark, cool, shed
A woodpile for chasing lizards
Lots of rabbits and cages
A hallway long enough for roller-skating
A rock fireplace with a place to sit

Homemade tortillas
Buckets of flour kept under the sink
Strong hands kneading endlessly
Humming forgotten words to a song
Balls of dough lined the table
A wooden rolling pin with no handles

I waited for the hot ones on the stove
Melting the peanut butter before I could eat it

Deer jerky hanging on the clothesline
Fried eggs and beans
My favorite was the sweet rice
Made with milk and sugar

Her big giant brown car
I was always glad to see it pull into the school parking lot
Especially the day I broke my arm
I didn't cry until I got in and shut the door

Christmas Eve

A huge event for the whole family
So many of us we could hardly fit into the room
Mountains of presents dwarfed the tree
The excitement was electric
Grandma and Grandpa on the couch in the corner
Throwing the bows over her head into a big pile
Everyone always remembers the bows

The house on the hill
The shed with the punching bag
The woodpile
The rabbits
Shep

So many memories
They should never be lost
We must share them
This way Grandma
And her guitar
And her tortillas
And her bows
Will live on

These stories, such as "Hands," are just a few of the many wonderful stories about Mama. Yes, she was truly spirited…yet in her latest years her gentle touch was like brush with a butterfly as she stroked your hair into place or straightened out your clothes-always being a mother, no matter the illness. We thank You, Lord for being able to experience the balance in this incredible woman who on her last moment on this earth, gathered her children and took them to another level of spiritualism as they prayed for her release from this world in the loving arms of beloved "Viejo." Her fiery spirit lives on in her children, grandchildren and great grandchildren. —Helen Raster

Hands

Strong hands, but loving hands. And it showed how much strength she had, even thougth she had a disease, her holding your hands was her of saying "I love you." She liked to hold things in her hands like cloth, dolls (Elizabeth-Tayler), & many other things. What I am trying to say is that I am grateful for her, especially her hands.

Writen by: Alex B. Raster
June 4th, 2001

Hunting Trip

Ben Luna

We always looked forward to going hunting with Dad and the family. We would plan the hunt almost a year in advance. We would pack our supplies in Dads old 1955 Ford truck and three to four of us would go all the way to Utah to go deer hunting.

We would stop in Las Vegas and register to win the prizes that the casinos would give away, and have a breakfast for 99 cents. Our next stop would be St. George, Utah, where we stopped in some large parking lot and slept for 3 to 4 hours until daybreak. Our destination was a small town called Koosharem, where we bought a few more supplies, then start up the mountain to our campsite on Greenwich Creek.

Hunting in Utah was always such a joy! The wilderness, the fresh air, the scenery was always spectacular. We would get up early in the morning, usually before daybreak, and go hunting for a few hours, and then we headed back to camp where Dad would make the best breakfast in the world, which consisted of fried potatoes, eggs and bacon.

At night, after supper, we would sit around the campfire and tell stories. Dad would tell us of his younger days when he worked in the Civilian Conservation Corps, a program that was started by the government to create jobs during the depression years.

Our hunting trips were always a great success in terms that we always had a good time, bonded with each other and had a chance to share something special with loved ones. Sometimes we would be successful in the hunt and bag a deer, but our experiences were something that I will never forget.

Buck Knives

Saleta Ann Luna

As a little girl I would look through my grandparents dresser drawers to see what "treasures" I could find. My grandfather had a drawer full of Buck knives he had received as presents over the years. I would pick out the knife I wanted for the day and would excitedly run up to him, "Grandpa, can I play with this?" He would look at me and reply, "These knives are very sharp and a lot smarter than a lot of people I know, so be careful, mi hita." As a mother now of two girls, Lydia and Lilliana, I can't even imagine my girls playing with anything like that!

Blue Eyes

Helen Raster

Cynthia Marie Gerardi was born April 21, 1963. She was the first in our brown-eyed family to have light eyes. She had huge blue eyes like the sky.

Blue Eyes—Cynthia Marie Gerardi.

When Patsy, her mother brought her to our house we were

all taken by those huge light blue eyes. Daddy *hit the nail on the head* when he bent close to her and looked deeply into her light eyes and he says, "I wonder if they can see as good as we can." We all busted up laughing. He put into words what we were all thinking.

In Him

Van Raster

Integrity, **N**oble, **H**umble, **I**ntelligent, **M**agnificent, Greatness and Honest. These are just a few words that I can use to describe this Magnificent human being that God lent us. My grandfather, Noberto G. Luna. I am not sure how other grandchildren feel about their grandparents. However, I do know this, I have been truly blessed to have such genuine rich-spirited grandparents.

The times I love the most with my grandfather is when we just simply sat under the trees in the shade in his yard. He usually sat on an overturned bucket of some sort…his three or four top buttons undone on his plaid shirt for comfort. Occasionally moving the water to another tree during the story sharing.

My grandfather did not speak rapidly, he spoke with clear concise well thought-of words before he spoke searching deep in his memory to be very specific. He would share stories about the past. Stories like the CCC Camps in New Mexico where he worked as a young man.

One of the things I really enjoyed is when I would visit him with my friends. I felt as if my friends were friends of his too. It was like a chief speaking in a tribe. Well-respected, full of wisdom, and nobody would talk, but rather listen closely because we knew nothing but the truth was being told.

Someone once told me, you could tell how rich-spirited a family is. When you have a family gathering when the families walk in with their children they will always go and pay their respects first to their grandparents. This is just the way it was and I loved it. "Waiting in line for the hugs"…*just as it should be.*

My Grandpa Luna

Stephanie Luna Gallop

My grandpa died almost a year ago.
I don't even have to close my eyes to see him.
His image is just there.
I can breathe him in.

I especially loved my grandpa's hair, silver satin and downy soft.
Long enough to touch his collar.
His cowboy hat sat carefully atop his shiny mane.
Long sideburns ran down to meet his unique mustache,
a mustache that curled up as if it were an extension of his smile.
And then there was his smile.
Not only on his lips, but also in his eyes.
Reflecting the adoration he had for his family.
And infinite understanding for even the smallest child.

His voice was soft, washing over my ears with soothing comfort.
It seemed he never wasted words, only saying what needed to be said.
His quiet voice claiming the attention of anyone present.

His hands were never idle
Years of throwing hay to the horses
Or tinkering with the abundant machinery
that often decorated the yard
But his skin was velvety smooth
as I placed a gentle kiss on his cheek.

His boots.
They seemed to tell his story.
Dusty and tired, from work or travel.
They had a life all their own.

I never saw him use a fork.

Warm tortillas were the only practical eating utensils.
He ate with slow deliberateness,
savoring his meal as if it were a glorious gift.
His life was a gift to us all.
He left behind a remarkable family.
A family who will never stop missing or remembering him.
The best family.
And I can still hear him say,
"Nothing but the best, mi hita!"

Grandpa Luna

Stephanie McCullough

I remember every time we went to California to visit, I would always go over to Grandpa Luna's house to ride his horse, Heidi. The first time I went over there was when I was six or seven. I had never ridden a horse before, so I was kind of scared, but I didn't show it. (I like trying new things.) Well, anyway, I rode her around the yard with Grandpa leading her. Then I got brave, and asked if I could ride by myself without him. So I did pretty well 'til Heidi stumbled on something and almost knocked me out of the saddle, I was scared spit less. Needless to say, I was very cautious about riding after that. After I rode her, Grandpa and I went to see the sheep, and the other horses. I would help him water them, and give them their hay and stuff. Then as we were getting ready to go back to the house, he would always chip us off some salt from the horse's salt block. I think that was the best part of the day, besides riding, even though I was scared senseless. But I'll always remember that. He promised to give us all the horse tack. As yet, I don't have my own horse, but I plan to get one eventually.

Grandpa Luna

by Matthew Raster

(printed as it was written)

Before Grandpa died ever weekend when my dad came down to Barstow we would eat breakfast at Grandpas house. He made the best eggs I ever had. Also the bacon was very good

Shannon Raster, left and Daniel Bryniarski right, showing their "garajes" (garages), showing where their baby teeth fell out.

Missing Teeth

I was seven years old when my first baby tooth started to wiggle. It took one week of wiggling for it to come out.

I was very exited to see how much money I would get from the tooth fairy.

The excitement of losing a tooth and getting money from the tooth Fairy was only a part of it for me. The best part was when I went up to my grandpa (Great grandpa) and proudly showed him my missing tooth. He would smile and point to the gap and say "Hey—— you got a garage.

by: Alex B. Raster
Date: March 14, 2001

Handwritten note by Alex Raster when she lost her first tooth at age 7.

too. {?}After breakfast we wold go look at the animals and Grandpa would tell us there names. My favorite horse was Heidi. We always used to ride him. That's all the best things I remember about Grandpa Luna. I miss him a lot.

Mi Familia

Michael Jerome Luna

My name was given to me by my mother and father, Benerito and Ann Eloise Luna—a name that I am very proud of. I have been teased about being named Jerome, but I have become quite proud of it. My understanding is that that was the name Dad wanted. My mom and dad are the greatest. I love to see them together. She is truly an incredible, beautiful woman. Her spirit is so youthful, and happy, and always loving and caring. Everyone who knows her, loves her. She always seemed to be so outgoing and free spirited. While I was younger and still in school, my friends would come over and later say "your mom is so cool." As I have grown older, I look to her, not only as my mom, but as a great friend as well. She has taught me so much about life and love and the joys that the lord has given us and how we are to cherish all them. Word cannot describe how I truly love my mother. When I see her, she always makes me smile. What a great feeling!

Dad is awesome. He is my champion. I have never known a harder working man. He was so wise and a pillar of strength within our family. I love him so much. There is nothing that he cannot do. When it comes to automotive knowledge, there is no one better. He is the best. He represents everything I want to be. As a father now, I recognize all that he did to provide a wonderful home filled with love and happiness for me and my brothers and sister. Today, I envy my brother being able to spend his days with him at the shop. Though I am far away, my thoughts are often with them both.

The Big House

As the oldest boy growing up in Barstow, with my younger brother, Tony, my sister Gina, and baby brother Stephan, I have only the fondest memories of being brought up and raised in an environment filled with family love and happiness. I remember Dad working two jobs, one at the Marine Base and the other at H & E Hardware, to be the best provider. Mom was a wonderful woman who took care of us and raised us. My recollections of that time only were of being a carefree kid, having fun. We always ate well and had a good home.

I can recall spending many days with my Nana and Nina, weekends and holidays with Grandma and Grandpa Luna at the "big house on the hill." As a kid, the house seemed huge. It had a hallway that was a mile long. On the nights that I would sleep over, I would be in the far off bedroom. I can still remember waking up early in the morning and finding Grandma in the sewing room. It would still be dark outside and there she would be, sewing away. The sewing machine was really old and made out of black steel.

My earliest recollections there were when my Uncle Randy still lived in the room second from the end. My cousins, myself and I would have pillow fights with him. Grandma was always making cakes for other people and she would always save the extra frosting for me. I remember going into the kitchen and finding the bowl filled with different colors of frosting in the refrigerator. She always knew I would find it. If it wasn't that, we were going for the fresh home made tortillas that she kept in the big chrome breadbox on the counter.

I remember at dinnertime, as we sat at the table, grandpa would tell me and my brother Tony, "If you guys don't finish your food, first the little one then the big one." He said this as he gestured to the window over the dining table. Tony and I would look at each other in wonder of what he meant. We later realized that he was saying that he would throw Tony through the window first and then he would throw me next. "First the little, then the big one." Of course he was kidding, but at the time, being ignorant little kids, we did just as he said.

Grandma and Grandpa's cooking was always great. Grandpa would always make us a great tasting breakfast. That is something that I just now realized that I do today. Every weekend, I make breakfast for my family. Perhaps it is by his influence.

Christmas was always fun there as well. All of the family

would gather in the den, with adults sitting on the furniture and all of the kids on the floor. Presents were piled all around the Christmas tree. We would open all of the presents at some super early hour. All of the wrappings were gathered and then burned in the big fireplace that was in the living room. They were times when all of the cousins would gather and have fun together. I grew up knowing and spending time with all of my first cousins. It was awesome. They were great times. Whenever we would gather at Grandma and Grandpas house, we would always be playing outside playing games with Shep or just getting into stuff. I remember being scared to go into the shack that grandpa had. There was a "speed bag" just outside the door. It was this tear shaped punching bag that grandpa would work out on occasionally. He was awesome to watch. I remember he would get that thing moving so fast with his fists, and then he would use his elbows. As kids, we would all just watch in awe. Neither one us could do it and all of us wanted to.

I remember Grandpa had all of his rifles and guns in this small corridor behind the fireplace in a gun case. As a boy growing up in the Luna family, I knew there would eventually come a day when I would be able to go on a hunting trip with Grandpa, Dad and the uncles. I remember Mom, Grandma and my aunts saying their goodbyes to the men as they departed for their deer hunting trip. It is still a tradition that is proudly carried on today. It is something that, I am very proud to continue with my sons.

The First Buck

On my first deer hunt with my new rifle, my dad, grandpa, Uncle Leo and myself, set out to hunt an area near the Mid Hills area in the Eastern Mojave Desert. I had been on hunting trips with Dad as a kid, but I think I was just walking the hills and hoping to just get a glimpse of any wild animal. This time I was seriously out to get a deer. I had spent hard earned money on a new Browning rifle, a fine weapon that met with Grandpa's approval. We had split up to cover the area and about one hour into the hunt, I had bagged my first three-point buck. I had bought a videotape of how to properly field dress an animal prior to the trip but had never actually even practiced on anything. I remember being very proud of what I had accomplished. The only thing missing from the moment was anyone to share it with. Dad, Uncle Leo and Grandpa had heard the shot and I knew they would be along soon. Out of breath, I knelt down and prayed to the Lord for the moment. As I stood up, I looked over to see my grandfather coming my way. I was so happy and proud. Still slightly out of breath, I called out to him" Grandpa, I need a coach." When he got there, he said" That's a nice one" as he shook my hand. I would like to think that he was very proud of me at that moment and for what I had achieved. Maybe he had had the same feeling in him as a younger man that I had in me at that very moment.

As the field dress was being completed by yours truly, Uncle Leo said that I had to drink the blood of the animal to gather his spirit. He said that it was tradition and that it was something that I had to do." Every hunter in the Luna family who had done this had followed the tradition." Of coarse I doubted him, but he kept on. "No, you have to drink it." Well, he persisted to the point where I knew I wasn't getting around it. Slowly, I commenced, they watched, all of them, even Dad…I did it! No sooner had I winced at the taste, did my uncle Leo burst out in laughter, saying he had never seen anyone ever do it. We all got a good laugh out of it. I will always remember that moment, with just me and my grandpa. I thank the Lord for that special time.

The Family

On a visit to Barstow one Christmas I met a beautiful woman named Anita Lorraine Saiz. We dated for five years before she became my wife. Sometime later, the first fourth generation Luna was born. My grandparents were very proud of our son Isaac Matthew. He was born on June 30, 1987. A photo was later taken of all four of us, Grandpa, Dad, myself and Isaac. They were just as proud when Vincent Anthony was born on August 30, 1989 and Adam Nicholas on May 24th, 1993. The Luna name lives on.

I have many more memories of my youth. I wish I had more time and the recollection to write about them. I guess they will be passed on, the old fashioned way. By spending good times with my aunts and uncles, cousins, nephews and nieces, sons and relations, recalling all the happiness of our past. My grandparents have left us so much to live for in each other. They have been incredible role models for us all.

And as we go on in life, and we think of them, they will always bring us a smile.

My Memories Of Grandpa And Grandma

Kristine Luna Bryniarski

Grandma took care of us after school so we would go right to the house on the hill. The smell of fresh tortillas right off the griddle made my mouth water. Grandma made them everyday it seemed. They were still piping hot and we would spread butter or peanut butter on them. They were the best tortillas and to this day I have never had any better than Grandma's. She never measured the ingredients, she knew by sight and years of experience how much to use. I would watch her knead the dough, adding what she needed to, then make little dough balls. Then she got out the rolling pin and rolled them out. With her left hand she would lift a corner of the dough flip it over her right wrist and lay it back down and roll. She did this until they were flat and almost perfectly round, then she heated them on the griddle. She made it look so easy, but when I tried it, my hands and arms got tired. And she made incredible sweet rice with cinnamon.

I also remember when my great-grandmother Barden died, Grandma was the one who told me, and I thought she was joking. She said she would never joke about something like that. Then at the funeral I cried so much that I had to sit at the back of the church. Grandma Luna sat with me and comforted me; she laid my head in her lap and stroked my hair as I cried like a baby.

Grandma had an old worn-out pink slipper. And when we did something wrong like slamming doors or one of our famous rock fights, Grandma would swat us with that floppy slipper. She had some muscle, that floppy slipper would sting the back of my thighs. She wasn't shy about using it, and I had the utmost respect for that slipper.

On Christmas Eve, all my uncles would take turns being Santa. They passed out all the presents while Grandma and Grandpa sat on the couch in the corner. Then as the grandsons got older, they would take turns playing Santa and hand out gifts. I treasure the memories I have of our entire family at Christmas time. And Grandma saved all the bows; she tossed them behind the sofa where she and Grandpa sat.

Grandpa came home from work, probably exhausted. We would climb all over him, sitting on his lap. Grandma would shoo us away saying that Grandpa was tired. He would respond by saying, "They're ok." He never seemed to mind spending time with his grandchildren no matter how tired he was. When I was in kindergarten, I put a small rock in my ear (ok- I was 5) and it got lodged. The school nurse couldn't get it out and sent me home. I remember Grandpa asking me how it got there and I responded, "I don't know." He laughed, and then he got tweezers and pulled it out. I distinctly remember showing Grandpa my "garage" when my two front teeth fell out and him smiling back at me and the sound of his chuckle.

The house on the hill was a huge house. It had the longest hallway I've ever seen (from a child's point of view). Behind the house and down the hill was a big woodpile where most of our rock fights took place. There was also the frame of a car down there that still had the steering wheel and foot pedals and we did lots of pretend driving there. There was no body to the car or seats. And never mind that there were tons of snakes and lizards that lived in the woodpile. It was a child's play land.

Another great memory is how our whole family would go on camping trips together, or pinion hunting in Big Bear, or Easter egg hunts on Easter. As a child I loved our family trips and as an adult I (emotionally) think about those times. I'll look at one of my son's mud covered faces and without thinking I say "pendejo." Then I laugh to myself because you know "you would of have to have been there'.

It was truly excellent growing up as a Luna grandchild. As I write this, I can't stop crying and laughing as the memories flood over me. I haven't thought of these memories in so many years. Where did the time go? I must savor every minute of my life and my boys' lives. I am now making the memories for my boys that I hope will some day bring them the same fond, cherished, intense, and hilàrious emotions that I have from my childhood. What a blessing being the granddaughter of Noberto and Annie Luna.

Memories Of Grandpa And Grandma

Cynthia McCullough

Grandma, my mom and Saleta came to see Lonnie and me when we were stationed in Germany. We took a trip to a few countries in Europe. We had a little yellow Toyota Corolla and we were packed in it like sardines, four adults and one little girl. It was crowded but we had so much fun. I remember one stop we made in Paris. We went to the Louvre, a museum that has many famous things of Europe, one of which was Napoleon's belongings. I could here Grandma mumbling something as she came to Napoleon's horse. I came over to where she was and asked her what she was talking about. She pointed to the horse (which looked more like a shaggy pony) and said something to the effect that she had a horse back home that looked better than that ugly horse and even Durango would look better in there than him. She then proceeded to talk bad about all his tack and saddle. We couldn't stop laughing. Here we were in France and Grandma is "dogging out" one of their heroes.

Another stop we made was in Switzerland. We went to the open market where they were selling a number of things. Grandma went over to the fruit and started picking some to buy. The merchant came over and tapped her hand and told her "no" and continued on in his own language making a fuss about Grandma touching the fruit (over there, they pick the fruit for you). All of a sudden Lonnie and I hear Grandma cursing and yelling at this man and with a few choice words tells this man he can take his fruit and shove it. My mom, Lonnie and I busted up laughing..

One of my favorite times of the week was Sunday. After we would get out of Mass we would all make a beeline for Grandpa and Grandma's house for leftovers from breakfast.

They would go to the Spanish Mass and we would attend the 11:00 Mass so they were finished with breakfast by the time we got there but they always made extra (I wonder why)? We would walk in the house and smell fried potatoes and chili. Sure enough we'd go straight to the kitchen and there they were: chili, tortillas, potatoes, beans and sometimes left over eggs. All we had to do was stoke up the fire and throw some eggs in Grandpa's special "egg" pan. Man, what a feast! I sure miss those times with them on Sunday morning. Grandpa loved cooking breakfast for everyone.

I remember one of the sayings Grandpa used to say that I use all the time now and can even "catch" some people like he did. When it was late and he was at someone's house getting ready to go home, he'd say, "Well, I better get going, 5:00 comes mighty early." And sometimes people would say, "You mean you get up at 5:00 in the morning?" and he'd say, " No, but 5:00 still comes mighty early." I love it.

I think Christmas was my favorite time of year when I was little. We would go to Grandpa and Grandma's house with the rest of all the families. The house would smell like all kinds of delicious food, the presents would be spread all around the tree and half of the family room, the place would be filled with laughter and loud talking, it was so wonderful. I remember hoping I was the first of the granddaughters to open a present from Grandma because I knew she made all of us the same thing and once one of the granddaughters opened their gift we all knew what we were getting. It was so special. She took so much time to sew or crochet each of us something. We all matched and I thought it was the greatest thing. She invested a lot of her time making special things like that for everyone and they were the best kinds of presents because they were hand made.

One year she crocheted snowmen for all of us, and this was when we were older and married. She was still making stuff for us even though we were grown and weren't there any longer. Well, my husband's snowman was tall and thin like him and she made my snowman short and round like me. I didn't like the thought of being short and round but it was sweet of her to take the time and make each of them to custom fit us. What a wonderful woman! I still have them and set them out each Christmas with such love and care and have crocheted the other two additions to my family so we are all there now. I have the little canvas house she made for tissue boxes still sitting on our toilet. I have a little of Grandma all over the place.

Being in the military, we have always lived away from home. But when we would go back to Barstow one of the high lights for my children was to go to Grandpa and Grandma's house and ride Heidi, grandpa's old mare, and gather Grandpa's eggs. Grandpa would get out there in the heat or cold and saddle Heidi for the kids and they would ride around and around the trailer and smile the whole time. Grandpa would then take his "egg basket" and give it to Caleb and he and Grandpa would go and collect eggs. He'd tell me about the latest happenings with his sheep, chickens, horses and rabbits. I loved to hear him tell of his fighting roosters and of course his prize stallion, Renal.

If you got there early enough there was always breakfast that he loved to make for everyone. I sure do miss the way he made his eggs.

Grandma taught me a lot of things that still continue on in my family. She taught me to play the guitar and we played together in the Spanish Mass and sometimes the 11:00 Mass. That was so much fun. She would sometimes play the bass guitar. How many people can say their grandma plays the bass guitar?

She gave me a love for music. I would remember being at the house and she would always hum or whistle to the radio. She would always hum the harmony not the melody like most people do. Hence, I started finding the harmony to all the songs I heard. Now, I love to sing, can't sing real well, but I love to sing and have passed that love for music on to my daughter who loves to sing. We harmonize to all the songs we sing. Our family sings in church all the time and I know that would just thrill Grandma's heart if she knew that it was because of her love for music that she passed on that started all this.

Grandma taught me a lot of things but one of the most valuable things is how to be a strong woman. She taught my mom who in turn, taught me, who in turn, is teaching my daughter. Women do not have to be weak and frail "no nothings." We can have the strength of an ox and yet be feminine. We can know how to fix things around the house and on the car without waiting for the "honey" to do the "honey do list" and putting it all on him. We can speak our mind (which we do more than we should) because we have a lot of good ideas too. We are not afraid to tackle a job that might be otherwise challenging or even impossible for the average woman. We can work along side our husbands and yet be tender enough to care for a crying baby. We can give our opinion in a discussion and yet be submissive and obedient wives. We can help in building our own house and still be able to sit down to the sewing machine, crochet or do needle work like Grandma did. She was an awesome woman and few can compare to her.

I thank God for letting me be raised in the family He put me in. We don't have a choice in who our family is but God chose to put me in the Luna Family and I thank Him all the time for that.

I look at Grandpa and see a man who is a pillar in society, a righteous man who loves and cares for others. Who did good and helped those around him. He had such a sense of humor. He was honest but not brutal (She's no Marilyn Monroe for a homely person). He was a good grandpa to me. I am one of those "huggy, kissy" people and even though he wasn't known for giving kisses and hugs a lot, after Grandma got sick, he would hug and kiss me so much more.

I miss Grandpa and Grandma very much but they are with the Lord living in complete happiness in a much better place than here and we will see them in heaven someday.

My Memories Of Grandpa And Grandma

Michelle Gerardi Jensen

When I was a little girl, Grandma taught me how to play guitar. It was so much fun sitting down with her to play De Colores and all of the other songs she played in the Spanish Mass. Eventually Grandma asked me if I wanted to play guitar with her in the Spanish Mass. What an honor! It was such a privilege to be able to stand up with Grandma and play, knowing she was the reason I was able to. She would call on Saturday night to make sure I was ready for Sunday morning. She and Grandpa would come pick me up about 6:30 a.m. so we could practice before Mass. Eventually I started playing guitar in the Spanish Mass, the 11:00 Mass and the 6:00 Mass. It was all because Grandma took the time to teach me. I can still hear her humming and singing De Colores. You always knew when Grandma was happy because she would be humming or whistling. If someone upset her, she would stop to say what she had to say, then go on about her business humming again as if nothing ever happened. The greatest times we had together for me were sitting down and playing guitar together. Later on, Grandma went on to play bass guitar in Mass. That goes to show you're never too old!

The last time we went home to visit Barstow, we went to the balloon festival in Albuquerque. That would be the last time we would get to see Grandma and Grandpa both. Grandma was just like a little baby girl. She would grab anything she could put her hands on. Jett, my youngest son, was loving Grandma one day when he had a painful, "help me" look on his face. Grandma had grabbed on to his hand and wouldn't let go. He was only two years old at that time. He just stood there until someone asked him what was wrong. He replied, "She's squeezing my hand and she won't let go!" What was a two year old to do? He had no idea how he was going to get Grandma to let go of him.

To this day, he still tells every one about the balloon festival and how Grandma wouldn't let go of him. While we were at the balloon festival, I spent a lot of time with Grandpa. We were both unable to walk around because the area was so large, so we sat in wheelchairs admiring the beauty of everything. We had such a nice visit. Grandpa was always worried about my health even if something was wrong with him. He never let anyone believe anything was ever wrong with him. The visit with him in Albuquerque would be the last one. Grandpa looked so sad and empty. I would hold his hand and ask him what was wrong and he would say, "Mija, I just can't handle seeing my *bieja* this way. I miss talking to her." He would look at her with such love in his eyes. He was so lonely. When it was time to leave from Albuquerque, I kissed Grandpa goodbye and felt as though it would indeed be the last time I got to see him. Grandpa was just not his usual happy self anymore. Two months later was when he went home to Heaven to be with our sweet Lord and Savior.

During that visit, all day long I would kiss Grandma on the cheek and say, "Hello pretty girl." Well, one evening when I was in my mom's kitchen visiting with Aunt Helen I reached over and said it one more time. Grandma was holding my hand. She looked right back up at me and said "Hello pretty girl" just as clear as could be. Grandma had not spoken to anyone. Aunt Helen and I just rejoiced and cried together, we never forget the look on Grandma's face, just as though she knew exactly who I was and what she was saying.

There is no love that can be shared like the love the Luna Family all have for each other. Grandpa and Grandma instilled in each of us the importance of a close, loving family. There is nothing more precious than love and that is something for Grandma and Grandpa to be proud of. Even in death, the family has continued on with the love and the memories left behind by both of them. We'll all miss them!

Memories By Jessica

From as far back as I can remember I have always felt blessed and proud of my name. It is unique and beautiful. My parents Nobie and Nora Luna named me Jessica Rene Luna at birth. I was born in Barstow in January 1970 and blessed to have both my parent's families living in Barstow. I remember thinking when I was five years old how wonderful it was that we all lived here together as a family. I have so many memories of my grandpa picking me up in his 1956 Ford pickup, and I would sing Spanish songs to him all the way home to the "Big House." When we got there Grandma would be cooking our dinner that usually consisted of beans, chili with meat, fresh tortillas, steak, and milk (also macaroni and cheese for us kids!). Grandma would hum a song the whole day long. A lot of times she would pull out her guitar and start playing a song and tell me and Saleta to sing *De Colores* with her, and we would happily oblige.

My childhood was the best: Before I started school my grandma would baby-sit us all day while my parents worked. On a typical day Saleta and I would eat breakfast together and then proceed to play and fight all day. On some days Grandma loaded us up in her 68 Ford Galaxie and drove us down to Sprouse-Reitz to buy some material for an outfit and we girls would sit on the stools and twist around and around. If we were good girls, Grandma would buy us a special treat...and at the end of the day Saleta and I would not want to leave each other (even if we did fight). I went home and spent a good time with my parents, got a good night's sleep and started all over again the next day! If I could have that back again for just one day...

My grandma was a tough one. She taught me that as a woman you did not have to be frail and you could speak your mind no matter. As strong, intelligent women, we bumped heads immediately! I am told by my Auntie Helen that I was no more than 1 ½ years old and Grandma was bathing me in her kitchen sink and I kept splashing water all over, she told me to stop—but to no avail. Finally she grabbed my hand and told me, "You stop splashing water!"

We glared at each other face to face with neither one of us budging. My aunt said if only she had a camera...she witnessed the birth of my awesome stubbornness that was inherited from my grandma—a whole generation apart!

I felt that I needed to carry on part of our families memories of Grandma's car so I decided to restore it back to the days when all of my family, at one time or another, rode in "grandma's car." So I asked Grandpa Luna and he agreed to it as long as I did not go fast in the car... because as you know that is a muscle car! And it was, we could hear Grandma coming down the road a mile away! Oh my word—"Did my grandma think she owned the whole road?" She always said, "As a matter a fact I do!" She could throw at least six of us in the car and we were off! That car was sweet, smooth and FAST!

I must tell you that I am so happy that I decided to restore the car! I have always been so very close to my family, but this was a different type of bond: my dad, grandpa and Uncle Leo would meet me at grandpa's house on the weekend and we would start to work. I was in awe of my grandpa. He was so intelligent and could go out to the car when we were having trouble and could name what he thought was wrong, and I'll be damned if he didn't hit it on the head. He knew every tool like the back of his hand. The conversations that we would have were so different because he looked at me as his granddaughter that was now a woman. He told me how proud he was of me that I was strong and smart. Do you know that meant the world to me? After working on the car all day we were talking about ages and he said, oh to be thirty again. I told him that I would love to trade places with him. I told him that he was the luckiest man I know. He asked me why? I said. "Grandpa, you are 83 years old, you are in good health, your wife is alive, you have this big family, and none of us have died in car wrecks, or died from health problems...we are all still here! Do you know how blessed you are?" He was silent for a moment and quietly said yes. That moment was such a powerful emotion we shared. After that I called to my grandpa and waited for him to look into my eyes and I told him, "I just want you to know how much I love you, Grandpa—thank you for my life."

He said, "I love you too, Mija."

And I thank Jesus every day for letting me tell him that, because the Lord took him home with him the next month.

Memories Of Grandpa And Grandma Luna

David Jensen

Back in 1995, when Michelle and I moved from Atlanta to Barstow, I hadn't yet met the Luna's. Any of them! Michelle insisted we go see the grandparents first. Grandma Luna was sitting in a chair in the living room. You could tell, even after Michelle had introduced me, that Grandma was still trying to figure out who the gringo was. She was glad to see Zak, our son, and fell in love with Haeleigh, our daughter, who was just turning three. We visited for about an hour. I was learning some great stories about the Luna family. Grandma was telling a story about a house they lived in while in Barstow. How the grandkids ran around in the yard, making a mess of everything. During one story, in particular, Grandpa Luna kept butting in. Grandma would be in mid sentence and Grandpa would break in, "No, no, no he was doing this..." or, "no, she was wearing that..." You could tell that Grandma was getting frustrated. Again in mid sentence, Grandpa interrupted again, to correct her on some misplaced facts.

Grandma had had enough! "Hey, you!," she said " you just shut up." She stammered for her words. "You just shut up, before I, before I...PUT YOU SOME EARRINGS! " All the while, poking a earring at him as if she were piercing his ears right there. All Michelle and I could was laugh. It still makes us laugh six years later. Grandpa Luna was someone I always felt close to. Having only known him a short while, the memories are few, but happy. Michelle was having a hospital stay in Riverside for knee surgery. Patsy was watching the kids for us, while I was with Michelle. I could count on going to pick the kids up and Patsy would have a meal on a plate, waiting for me to get there. On this particular day, I arrived to get the kids just as dinner was being served. Grandpa Luna was there, as he was on most nights. Patsy made him a bowl of chili, and not the wimpy kind that I eat. I mean, it would make your eyes water just smelling it. You'd look over at Grandpa Luna, and he'd be eating it like there was no tomorrow, a small bead of sweat forming on his brow. I couldn't tell if it was because of the warm air, or the hot chili. We were chatting and carrying on. Grandpa asked me if I would like to try some of his chili. Well, me being ignorant of how spicy the chili was, and how Grandpa knew EXACTLY what he was doing, just had to have some. Before finishing my first couple of bites, I was in trouble. And Grandpa Luna was giggling like a kid in a toy store. "What's the matter, David, You don't like the hot chili, mijo?" and the voice trailed off in laughter. Those are just a couple of memories that stick out, for Michelle and me.

My Most Profound Moments

Dena Gerardi

One of the most profound and powerful experiences I've had in my lifetime centers on the patriarch of the Luna family, my husband's grandfather, Noberto Luna.

In 1995 when my husband asked me to marry him, he informed me that we would have to go visit his grandparents, Noberto and Annie Luna, to ask for his grandfather's blessing of our impending marriage.

Up to this point my exposure to Mark's grandparents had been minimal, limited to a few brief encounters. I didn't know them very well. Needless to say I was terrified of Grandpa Luna and hoped that he would approve.

We set up a date and time to go visit Mark's grandparents. I baked an apple pie in offering, "the way to a man's heart is through his stomach"!? Off we went for our blessing visit. I was a nervous wreck. My heart was pounding.

When we arrived we made small talk and watched Lente Loco or something like it on the Spanish channel. By this time Grandma Annie had advanced Alzheimer's. She said very little and was not able to manage in the kitchen alone. Mark suggested we go into the kitchen and fix pie and coffee for everyone. Grandma Annie and I went into the kitchen leaving the men to talk. Grandma Annie began taking out every glass and cup from the cupboards and filling them all with milk. I took the milk and offered to help. As I began pouring the milk back into the jug, Annie went after the pie. She took a big knife and began to hack the pie into tiny bits. I was terrified and didn't know what to do. I was so afraid she would hurt herself with the knife (or me, the blonde stranger intruding in her kitchen; of course I had heard all the stories about how feisty she was, but had never witnessed it!). So I told her to go relax in her chair and I would serve her. Thankfully, she obliged and I proceeded to serve the mangled pie to all. When I gave the pie to Mark he just smiled as he had seen what was going on in the kitchen. Grandpa looked at the pie and declared "It looks good." I could tell he was wondering what kind of cook Mark had found for himself.

When we were all seated eating the mangled pie, Mark told his Grandpa that he had asked me to marry him and we were here to ask for his approval and blessing. Grandpa appeared to give this serious thought. He then stood up and took Mark's and me by the hand. We all stood in a circle in the middle of the living room holding hands (Grandma Annie was busy with her pie and watching her program.) Grandpa began a little prayer and blessed our union. I was so overcome by this I couldn't not hold back the tears. He then made the sign of the cross on Mark and hugged him tight. He turned to me and did the same but he took my hand and kissed my cheek. My knees actually buckled! This was one of the most powerful and special moments of my life. For that brief moment I got a glimpse into the heart of a man that I had yet to know but grew to love and respect and care for. At that moment I truly became a part of the Luna family.

From Grandpa Luna down to the littlest cousin, the Luna family has always made me feel loved and welcomed with open arms. I am blessed and proud to part of the Luna clan.

On a personal note: I miss Grandma and Grandpa Luna, but my heart is warmed because I know they are with Our Father and together in each others arms once again, happy and free. The legacy of love and family they created and passed on remains with us.

Annie Z. Luna
1918 - 2001

- She never finished high school,
  But she excelled at the complicated geometry of raising six of her own children and one granddaughter.

Annie Zamora Luna, the feisty matriarch of the Luna family.

- She never got a degree in engineering,
  But she was an outstanding domestic engineer.
- She didn't know Algebra or Calculus,

But she could make a dollar stretch enough to feed her family.

- She didn't go to medical school,

But she kept us all reasonably healthy.

- She didn't go to chef's college,

But she made the most wonderful meals.

- She never took a single music lesson,

But she could play the harmonica, bass guitar, lead guitar, the organ, and the concertina.

- She never taught school,

But she taught her children to sew and cook and her grandchildren how to play guitar.

- She played in the church choir even though she never had a lesson in her life.

Our mama was diagnosed with Alzheimer's Disease in March of 1994. We saw her go from an energetic, resourceful woman to a shell of her former self. The dreaded disease finally won on March 25, 2001 when she took her last breath.

She outlived her husband of sixty-one years by fifteen months even though she had no idea he had died. She left behind six children, twenty grandchildren and thirty-six great grandchildren. She was a woman with such spirit that it will live forever in her children and grandchildren, but she will be missed by all her family and friends.

This was Annie Z. Luna. This was our mama, grandma, great-grandma.

The Final Story

Helen Raster

To the reader these pages may seem somewhat like a eulogy. I believe for all those who shared their stories and memories in this book, it ***is*** a Eulogy. Between the conception of this book and the completion, we lost our mother. We were already affected by the sudden loss of our father, when the subject of this book came up. The Cinco De Mayo presentation at the museum on May 5, 2001, was on his birthday. A date we can't shrug our shoulders and coin as a ***coincidence***.

In the preparation for this book, we have all experienced a sense of the past; a reaching back into time to document births, deaths, marriages, relocating. We all took our parents for granted, never realizing the sacrifices and the courage it took to move a family to a different state and to raise a large family. Losing parents has been very unsettling for us. We have felt a disconnection with the very people we are married to, and our own children cannot fill the void.

For some of us, the journey to finding ***who we are* and *where are we going*** is almost over, but for others the road has yet to be traveled.

One thing for certain, our parents are definitely ***where we came from*** and ***who we are*** today.

Esther Arredondo

Part 5
The Arredondo Family of the Mojave Desert

The Arredondo Family in Hinkley in 1936. Front row, L. to R., Ramon, Annie and Ruben. Back row, L. to R., Fela, Paula, Aunt Lupe, Jose, Hermenia, Elvira and Minnie.

Mrs Penny
Best Wishes 11/18/01
Ruben "Beny" Arredondo

Ramon Arredondo Family
Paula Arredondo Langbar

THE ARREDONDO FAMILY

Jose Arredondo

When Jose Arredondo was 12, he lost his father, Raphael, suddenly to typhoid fever. His passing left the family without a breadwinner as Raphael ran the "rancho" selling livestock, mostly hogs, and raising alfalfa as a cash crop. His mother, Paula Contreras de Arredondo, could not manage the farm alone and her children were too young to help. Raphael had saved a lot of money, mostly in gold, unfortunately he distrusted banks, and buried it for safekeeping but had not told his wife where. While delirious on his deathbed, he tried to tell his wife something but she could not understand him. After he died the family looked for the money without success. A couple years after the ranch had been sold to distant relatives, they found the gold buried deep under the hog pen; however, they tried keeping it secret and would not share any of it with Rafael's family.

After selling their ranch, Paula moved her family to Irapuato, Mexico, where Jose went to school for a few years. While Jose was a good student, loved to read and studied geography and history, he decided it best to quit school and work in order to help support his family. He joined his two cousins, Jose and Jesus Arredondo, in selling material and "rebozos" (beautiful bright colored shawls) that were carried on the backs of donkeys. The young vendors traveled through small towns and villages in the State of Guanajuanto and neighboring State of San Luis Potosi for four to six weeks at a time, returning to Irapuato for a week or so. Jose explained to his son Rueben that the three of them learned a lot about people, especially about women and life during their early years of salesmanship. They sold enough to help with families with day-to-day living.

Jose Travels to the United States

Jose Contreras Arredondo was 17 when he came to the United States in 1898. Jose traveled with his second cousin Jesus Arredondo. Jesus was a very young man of about 15 or 16 years old. The two came to explore and work in a new country in search of a better way of life. Jose worked for a number of

Seated, brother Miguel and Mother Paula. Rear, sister Guadalupe and Jose Arredondo at Amboy in 1923.

railroads including the Missouri-Kansas and Texas, Illinois Central, Missouri Pacific, El Paso and Santa Fe railroads. For a number of years his work kept him in the Southwest and throughout the Midwest. The workers were made up of large gangs that laid new steel track. At that time, the gangs consisted of predominantly Irish Immigrants. Jose was fortunate to have a good friend of Irish and Mexican descent who recommend him to his general foreman to because he said Jose was also part Irish! Jose worked very hard for the next 12 years on the various railroads and would send money home to his family.

Jesus, his cousin, returned to Mexico a year or two later and joined his brother Jose in San Luis Potosi where they started a clothing business. Eventually they controlled the sales of the famous "rebozos de Santa Maria." These rebozos, were and still are, hand-made from pure silk and are still popular as evening wraps. The brothers opened several stores and set up all their grown children with businesses of their own. The Arredondos from San Luis are quite wealthy and well known in that area.

Jose Contreras Arredondo took the surname of Arredondo that was his mother's maiden name because his American fellow workers said it was easier to pronounce than Contreras. In Latin countries both the mother and father's name is used together. The name Arredondo originally came from the Northern Coastal province of Santander, Spain, and the name Contreras comes from the Kingdom of Castile in the area and city of Burgos. Jose's father's mother, Maria Delores Contreras, spoke Hebrew and owned several books in printed Hebrew but he had very little information about her background.

Jose Meets Herminia

Jose met Herminia Hernandez in Santa Rita, New Mexico, and they were married a few months later in El Paso, Texas, in 1910. Jose had accumulated a nice nest egg from saving his hard-earned railroad wages and occasional poker winnings, enabling the newlyweds to enjoy a three-month-long honeymoon in Irapuato, Mexico (Jose's home), Guanajuato, the state capital, and Mexico City.

While Jose and Herminia were still on their honeymoon, the Mexican Revolution began which lasted five long years. Realizing they would be unable to leave easily and because he longed to return to ranch life raising animals and growing crops in the rich soil and ideal climatic conditions of Central Mexico, Jose bought a small ranch. In addition to Jose and Herminia, Jose's mother, Paula, his sister, Guadalupe, and his half brother, Miguel, lived with them on the ranch raising alfalfa, peanuts, hogs, horses and most of their fruits and vegetables. Their life was good and they had their first two children, Elvira and Jessie at their ranch, "La Calera."

Sitting, great, great grandfather Jesus Hernandez and grandfather Alejandro Hernandez, Herminia's father.

The Mexican Revolution

As the Revolution continued Jose realized he could no longer stay on his ranch, neither protecting his family nor staying out of the war. He joined General Venustiano Carranza in 1912, serving as Captain from 1912 to 1915, commanding a force of 60 men in the Irapuato-Guanajuato area. He fought in several battles under this Army General who became the President of Mexico in 1915.

As Jose feared, in his absence, the family's stock and crops were lost over the next three years to warring fractions. Jose became disillusioned and disgusted with all the terrible destruction the war was causing to human life, animal life and the land. The deciding factor that brought him back to the United States was his mother's terrible experience with a bunch of raiding soldiers (probably from Pancho Villa). While Jose was away fighting in the army, his mother Paula and brother Miguel managed the ranch. All of their animals had been confiscated in the war except for one workhorse used to plough the fields and one baby colt. On one frightful day, raiding soldiers stopped at their ranch to take their animals. The horse and colt had been hidden in a protected ravine. When Paula told them they no longer had any animals, the sergeant did not believe her, as he knew many ranchers were hiding their livestock. An argument ensued with the sergeant that escalated, and in her anger Paula accused them of being cowards taking advantage of women and children. The sergeant dismounted, walked over to Jose's mother and told her, "Don't think I won't hang you from that nearby alamo tree as one more hanging is nothing to me," he said. The family members were so terrified; they literally tried to push Paula into the house. However, she stood there glaring at the sergeant and said, "I don't doubt you would as it does not take a brave man to hang an old helpless woman." The sergeant's troops were as nervous as Paula's family as to what would happen. Suddenly the sergeant mounted and yelled his men, "We have to go now but we'll be back someday." Jose's family knew he was very serious in his threat and that they may not be as fortunate next time.

In 1915, Jose secretly made plans to escape the ravages of

Mexico by returning to the States. He wrote a former friend and boss from his years of laying tracks, asking the Santa Fe Road Master for help in obtaining a position as Section Foreman in the Southwest area. In the meantime, Pancho Villa's men happened upon his ranch when he was there visiting. He was forced to join them or be shot. As Jose rode away, he was already formulating a plan to escape. His first opportunity came the night before the next big payday. The men in camp were already drinking and carrying on, so Jose was able to quietly ride out. As Jose had anticipated, nobody thought anything about his absence the next day. After all who would believe that a soldier would leave without collecting his pay?

Riding back to his ranch as fast as possible, collecting his wife, his two children and a few possessions, they caught the train to El Paso where Herminia's family was waiting for them. Jose had a letter from his father-in-law stating his wife was in bad health and that it was important to visit her. If needed, this letter would be presented to Mexican officials upon entering El Paso. The train to El Paso was carrying troops and Jose recognized one of Villa's men from the camp he had just escaped from. Thinking quickly (before being spotted), Jose pulled his hat down upon his face and held his baby, Jessie, the rest of the way in front of his face or holding him on his lap so he could keep his head down. The train ride was dangerous but they arrived safely in El Paso without incident. Later that same year, he was appointed Section Foreman of the Ludlow Station in the Mojave Desert about 53 miles east of Barstow. His brother Miguel, sister and mother stayed in Mexico until the ranch sold and then joined Jose and his family in Ludlow. On their way to join Jose, his mother and sister crossed the border in El Paso with all their belongings. They had more than they could handle comfortably on their own, so they entrusted some of their belongings with a friend to store, including the family books in Hebrew. Unfortunately, these books later "disappeared."

Herminia Hernandez Arredondo

Herminia Hernandez's family came to Central California when she was just two years old in 1894. Her father, Alejandro, was an educated man who had taught school in San Luis Potosi, Mexico. Alejandro, hearing about the mining jobs in California, decided to move there to obtain a better life for his family. He labored in the mines and also did clerical work for the foreman of the mines. His work took him to many areas in California, such as Marysville, New Idira, Mariposa, Coalinga, and Hollister. Herminia was always proud of the fact that she came to California at such an early age, and many times made a joke that she was a "Californian," not a Mexican. Herminia was born in Tacubaia,

Misaela Abrego Hernandez, Herminia's mother.

a suburb of Mexico City, and did not have any memories of her childhood there. Perhaps that is why she always thought of herself as a native Californian first and a Mexican second!

With the continual moving around to the mining communities, Herminia's education was sporadic and thus she only attended school for about 2-3 years. This was enough for her to read and write in Spanish. Later she taught herself to read and write in English. Life was not easy in those years; the constant moves of her father made income an on and off situation. She would often say that the central part of California was very beautiful with the streams, mountains and wildflowers, especially around the Marysville and Mariposa areas.

In about 1906, Herminia's family moved to New Mexico then to Texas where her father worked for the mines and railroads. Herminia worked for a doctor of German descent in El Paso as a housekeeper and babysitter. She learned much about German customs, culture and cooking. Herminia became diversified in her cooking skills and in her sewing ability. She was an outstanding cook not only in Mexican cuisine, but also in American and German cuisine. She was known for her yeast breads, fried chicken, enchiladas and lemon meringue pie.

Herminia became Americanized in her thinking and her beliefs at a young age. She was a strong-willed woman who believed that women should have the same rights as men. (If she had been born thirty years later she probably would have been protesting for women's rights in San Francisco.) Her beliefs for women's rights were way ahead of the times and passed on to her daughters as adults. Herminia would not hesitate to correct sales people in stores or any other situation where she felt she was not being treated respectfully. Herminia had some memorable arguments with brakemen and railroad conductors. The family usually traveled by train because the Santa Fe provided free passes to their employees. Herminia would not stand for any nonsense or discriminatory acts when she boarded a train. If a conductor did not put down the stool on the ground to help her step up to the coach car, she would ask for one and not board the train until it was placed there. She would then give the conductor a good tongue lashing down the aisles of the coach for being so insensitive to women passengers.

Herminia's Family

Herminia's father, Alejandro Hernandez, came from the State of San Luis Potosi, and was considered quite well educated with beautiful penmanship and excellent command of the Spanish language. He became a schoolteacher at a young age, but Alejandro had itchy feet, liked to change jobs and do many things. He traveled extensively throughout the States of San Luis Potosi, Nuevo Leon and Guanajuanto. However, when he was teaching he also participated in military training and belonged to a well-

Mother Herminia, Elvira and Aunt Lupe.

known unit called "Los Carabineros de Nuevo Leon." This unit was well known for their shooting skills and horsemanship abilities. Alejandro was somewhat of a character. He liked his tequila and women and became quite well known in his travels as quite the womanizer.

Alejandro's cousin, Vicente Hernandez, was also a schoolteacher and assistant to a Hacienda Administrator in the State of San Luis Potosi. The two cousins taught school together for a while in a huge hacienda. Some haciendas were huge, having many employees called "peones"—thus they had "peones" and craftsman of various types to enable the haciendas to function as small townships.

Alejandro had another cousin, Maria Hernandez Mora, who visited Herminia and Jose when they lived in Hinkley, then later in Barstow. She liked to recount her many family stories. In the evenings, Ruben remembers that they would sit around the large kitchen table listening to her many exploits of traveling by horse buggy and of her life as a youth in the Haciendas. She would talk for several hours and most of the time fortified her long stories by smoking cigarettes and having a few shots of tequila or whiskey. Jose and Herminia did not allow hard liquor in the house but an exception was always made for Maria.

Herminia's mother, Michaela Abrego, married Alejandro at a young age in the State of San Luis Potosi, traveled and moved frequently with her husband. Michaela's parents came from Spain; unfortunately very little information is known about them.

Herminia was the oldest of three sisters and one brother who lived in the Los Angeles area except Alejandrina who lived in Phoenix, Arizona. The Arredondos frequently traveled to Los Angeles to visit their Aunt Maria, Aunt Enriquetta (Queta) and Uncle Jessie. The families remained close until their passing. Aunt Queta still lives in Pasadena and is 86 years old as of 2001. The children of the families still keep in touch.

Jose And Herminia's Return To The U.S.

In 1915 Jose, Herminia and their two children, Elvira and Jesse, moved to Ludlow where Jose was appointed Section Foreman for the Santa Fe Railroad. At that time Ludlow was a thriving community with approximately 400 to 500 residents. Ludlow was an important point on the main railroad line between Los Angeles and Chicago because here is where steam engines took on water, picked up and dropped off supplies and mining crews for the mines in that area and north to Eastern California along the Tonapah and Tidewater Railroad.

Ludlow had two general hardware stores, one of which was owned by a Chinese family named Lee that sold all types of hardware, groceries and dry goods. The town had a train depot, freight house and pumping station. Tank cars from Newberry Springs and other stations that had good water wells delivered water to Ludlow. Santa Fe provided water not only to the employees but also to the rest of the town. Ludlow had a Constable who provided law enforcement to the many small stations in that part of the Mojave Desert. The Tonopah and Tidewater Railroad ran north from Ludlow through Baker and Shoshone through small mining towns of Nevada toward Tonopah, a large mining town. Ludlow was a lively, busy community because of all the railroad employees and mining families.

Santa Fe Employees

The Santa Fe Railroad provided housing for the railroad crew. The Section Foreman and the Railroad Agent lived in wood frame houses that had basements and double roofs. The double roofs gave extra protection against the hot sun. There was approximately two feet of space between the roofs providing insulation to the house in both summer and winter. These homes were well built and offered a decent place to live in the desert. The track workers lived in long concrete or wood bunkhouses. Each bunkhouse had one or two rooms for the entire family. Some of the larger Section Stations had up to 15 track workers, depending on the location. Santa Fe provided old railroad ties or coal to their employees for heating and cooking. Ice, precious ice, was also provided to the workers. One hundred-pound blocks of ice were used in the ice storage boxes for perishables. Most of the homes had cellars to keep fresh fruit and vegetables. Families also used the cellars to cool off. Life was tough, wages were low and there was little entertainment and leisure. The families had to contend with the intense desert summer heat four to five months of the year. There were no fans or coolers at that time and temperatures would escalate as high as 120 degree F – sometimes even higher!

Section Gang Workers

Section gang workers in the Mojave Desert were a tough resilient bunch (speaking specifically from about 1900 to the middle 1950s). They labored without hardly any mechanical tools that are used in the industry today. Weather conditions were severe especially in the Amboy, Bollo, Cadiz and Needles areas. The summers were intensely hot, winters could be very cold and there were fierce windstorms throughout the year, making their jobs even rougher. Many men, even back then, would not have been able to endure the harsh weather or horrible working conditions. These men lived quietly with their families, tended to their jobs day in and day out, hardly complaining. Some of the men drank hard like other hardy laborers of that era. These desert gandydancers had a work ethic that was admirable and asked no special favors from anyone at anytime.

It is interesting to note that these gandydancers have had little or no exposure by books or television from an American historical perspective as has other groups of workers. All they did was to ensure the Santa Fe Railway that their trains would run on schedule, day in and day out, even during times of extremely heavy troop train traffic during World War II. In that note, we must mention that during both wars (WW I & WW II), their loyalty and work dedication was most essential to our country winning both world wars. It is our opinion that our nation owes these desert workers and their families a mammoth thanks and gratitude for their contribution to ensure democracy in our country and in the free world. The Arredondos are honored to be able to give these workers some recognition while writing their family history.

From about 1915 to the early 1940s most of the track workers, working in these isolated areas, were either Mexican or Mexican descent. Ruben's observation and comment is that, "When people live a hard, simple life in isolated areas with limited resources, the human spirit and heart work more in harmony to inspire people to join together making up for the lack and of these resources. They become closely bonded to ease those negative aspects of life that faced the desert track workers and their families in the 1920s, 1930s, and 1940s." When a family had an abundance of food, it was shared with the other workers usually this was when a pig or cow was butchered.

Desert Social Life

The Arredondos became close friends with other families from the outlying stations such as Newberry, Hector, Bolo, Amboy, Cadiz and Goffs. They became "compadres" to many friends at those stations. Compadres and Commadres are godparents of a baby or act as the best man or matron of honor at weddings bonding them. Families became so bonded or united that they all participated when there was a baptism, anniversary or birthday. At these gatherings somebody would strum a guitar, violin or mandolin, making it even more festive with dancing. These celebrations were usually held outdoors.

These people lived a simple isolated life style with limited resources; however, their spirit was such to enjoy one another, their families and to make the most of life.

Jose had few vices as he rarely drank and never smoked, but he loved to play poker. He learned to play cards while working in the gangs who laid steel track in Oklahoma and Kansas. The Ludlow poker games were held at one of the back rooms of one of the stores. Most of the players were railroad workers and some local store and business owners who were Mexican, Chinese or Anglos. Anyone who had enough money could play. The Ludlow Constable was known to crash a game, announce that gambling was illegal and that they were subject to be penalized. He would then break out in loud laugh and pull out a bottle of whiskey. The penalty was that each player had to drink two swigs of his whiskey per player. Jose hated the taste of whiskey, but he would always comply, drinking a cup of sweetened coffee as a chaser. The Constable usually joined the game with his whiskey bottle close at hand. Sometimes these poker pots would be large, as much as $200 to $400. Trainmen often carried considerable sums of money, especially on paydays. The Chinese often made sandwiches and coffee for the players for tips. It was not uncommon for these games to go on all night, breaking up before it was time to go to work. Jose was a good player and many times won some big pots.

Jose was a thrifty man and managed to save enough money from poker games to buy a ten-acre farm in Visalia located in Central California. The farm had a house, many fruit trees and grapevines. In the hot summer months Jose sent all his family there. Herminia loved Visalia and mentioned several times that it was like living in paradise compared to Amboy or Cadiz. Their third child, Paula, was born there.

Bolo

After living in Ludlow for a year Jose and Herminia were sent to the Bolo Station. Bolo, 11 miles east of Amboy, was smaller than Ludlow with nothing more than a Section Foreman's house and a bunkhouse for the crew. There were only a few trees and Hermina thought it depressing to live in such an ugly and out of the way place. At least in Amboy and Ludlow there were a few stores and businesses and many people with whom to socialize. Jose bought his first car when they lived in Bolo and managed to learn to drive, so they could venture to Amboy to shop and visit friends. Elvira remembered that the family dressed up for their weekend drive for cold Coca Colas at Roy's Café in Amboy. Her father wore a long duster and goggles (like in the movies) and sometimes put a flower in the flower holder by the window. About this time, Jose's family joined them providing more companionship for Herminia. Miguel, who was older than Jose, joined the trackers, but he did not like having Jose as his boss.

Cadiz

The following year the entire family moved to another section station further east known as Cadiz. The railroad consisted of the usual foreman house, workers' bunkhouse, and depot and a couple of other company houses lived in by the agent and telegraphers. Two miles away along the highway was a cluster of homes, a store and later café and gas station that served the increasing traffic on "Route 66." The railroads issued "passes" to their employees (and families) giving them free train travel. Jose and his family depended mostly on trains for their personal travel.

Because of the lack of grocery stores in the smaller stations like Bolo and Cadiz, the Santa Fe Railroad operated a company store called the Holmes Supply Co. for their employees. Workers could buy from the store on credit and then that amount was deducted from their next paycheck. This store operated much the same way as the company stores in the Appalachian coal mines in the eastern part of the States.

When trackers were first hired, they were given credit to buy groceries, clothing, shoes, blankets and any other goods needed to start housekeeping. Of course these prices were high so that on payday the trackers received little money from their paychecks. Some workers were in hock all the time to the Holmes Supply Co. Although akin to indentured labor, this system provided the worker a job and means to care for his family and also provided a better lifestyle than many Mexicans had in Mexico—hence the immigration.

Jose was not only frugal but he was also shrewd. He caught on quickly to the pitfall of buying from Holmes. Determined not to be in hock he began ordering from Ralph's Market in Los Angeles. They delivered his order by train within five to six days and by his ordering in large quantities the family saved even more money. Trains were the lifelines for the people of the section stations by providing transportation, delivering mail and shipping food and other supplies from Los Angeles, San Bernardino, Barstow and Needles.

Amboy

In 1919 the Arredondos moved to Amboy, which was similar to Ludlow, only hotter. The Santa Fe had a depot, freight station, section houses, a couple of stores and a schoolhouse. About eight or nine miles out of town there was a large salt mine that employed about 80-90 workers, which still exists today. The population of Amboy was between 200 and 300 people.

Jose and Herminia always said that Amboy had to be the

L. to R., son Jesse, Jose Arredondo and daughter Elvira at Amboy in 1925.

hottest pace in the Mojave along with Needles. Herminia and Guadalupe (Lupe) started washing outdoors as early as 4:00 a.m. before the heat. Cooking for the entire day started at 5:00 a.m. and by 7:00, breakfast; huge stacks of homemade flour tortillas and beans were made as well as some kind of meat and potatoes for their lunch and dinner. By noon their cleaning, washing and cooking was done for the day. After lunch the family, except Jose and Miquel, retired to the cellar for a nap on the their cots. Sometimes they also made fresh orange or lemon juice or do some mending to pass the hottest part of the day there.

Jose and Herminia's fourth child Herminia (Minnie) was born while they lived in Amboy. Minnie was the only baby born at a station. Usually, Herminia went to her parents or her sister Maria to have her babies delivered by a doctor. Herminia's sister-in-law, Lupe, stayed behind to take care of the other children and cook. A couple years after Minnie was born, Rafaela was born in El Paso where Herminia's folks lived. Again, Aunt Lupe took care of the children and home. Jose and Herminia still lived in Amboy when their younger two children, Annie and Ramon, were born. By the time their youngest, Benny, was born they had moved to Hinkley. On those three births Herminia delivered in the Santa Fe Hospital in Los Angeles and afterwards stayed with her family until she and her baby were strong enough to ride the train back to Hinkley. In those days an expectant mother was always in fear that she might not make it to the hospital and give birth right there on the train. That actually would happen from time to time and once to one of Herminia's commadres.

Jose and Herminia were close friends with Mrs. Starkey, the Amboy Elementary School teacher who taught all the grades in one schoolhouse. She was an excellent teacher, loving and

Paula in Los Angeles while in college in 1937.

kind to her students. Elvira, Jessie and Paula attended this school where the ethics of studying hard was instilled upon them. Mrs. Starkey was dedicated to her profession who believed that through a lot of hard work and being an honest person, you would succeed in whatever endeavor you pursued. Her philosophy paid off for Elvira and Paula Arredondo. They became excellent students and Paula was top of her classes throughout high school and college. While Jesse did not aspire to the academics, he continued onto the Los Angeles Technical School for automotive training where he graduated before he was eighteen.

Amboy Celebration

Some social events were segregated between the Anglos and Mexicans and some were not. Later when the Arredondo family lived in Hinkley in the 1930s and early 1940s, the Barstow Theatre had a policy stating where the Mexicans could sit, on the left hand side. However, Ramon and Ruben just sat where they wanted. "Nobody ever told me or Ramon where we could sit.

Holidays, like the 4th of July, were celebrated by all the townspeople as well as school plays and programs. One year the Mexican people of Amboy decided to celebrate Mexican Independence Day, September 16th, in grand style. They planned the celebration months before and the newly formed committee selected Jose Arredondo to be the chairman. A queen was selected to reign over the event and food (pork, beef, tortillas and chilies) was ordered in advance from Holmes Supply and Ralph's. The committee had an official program of the event printed in red and green, decorated with Mexican flags. There were patriotic speeches made by local people as well as guests from Ludlow, Needles and several other small towns (stations). Jose was one of the main speakers. All the people of Amboy and guests celebrated with much gusto especially at the Saturday night dance that went on to the wee hours. The celebration itself lasted all weekend. The fiesta area was gaily decorated with red and green crepe paper streamers were so festive that they were left up for a week or two.

In the middle of the 1960s, a lady by the last name of Flores, who had lived in Amboy at that time, still had one of the

Jose Arredondo and family when they lived in the Section Foreman's house in Hinkley in 1937.

colorful programs. She told Ruben Arredondo that she had known the Arredondos quite well and still remembered the celebration vividly. Although that program was rather worn, Ruben regretted not making a copy. It would have been quite appropriate to have it on display in the Mojave River Valley Museum in Barstow.

Hinkley

Santa Fe transferred Jose Arredondo to Section Foreman of Hinkley in 1929, which was a thriving small farm community with a general store, auto garage/station and a post office. This way his older children could go to Barstow Union High School .The post office and store, known as Burns Red and White General Store, were in the same building. The Santa Fe Depot had a small but picturesque depot with a freight area, ticket counter and waiting room. There was also a large freight house with a loading dock used by the farmers in transporting supplies, tools, and farm equipment. There was a small stockyard for transporting cattle and sheep that was often used until the 1950s.

Ramon and Annie at the railroad Section Foreman's house in Hinkley in 1944.

The Railroad Agent's house was next to the depot and across the tracks were two small houses for telegraphers. The section foreman's house and section bunkhouse were west of the depot on the same side of the tracks. Between the depot and foreman's house was a fenced small park area with grass and many cotton-

Hinkley railroad station in 1948 or 1949.

wood and mesquite trees. The original foreman's house was old, but in 1933 the railroad built a new larger house with a nice big front porch and concrete basement. The house was modern unlike the other railroad houses Santa Fe provided. The basement was well built and even in the hottest days the temperature never went over 72 degrees F. The family congregated there in the hot afternoons where they enjoyed ice cold watermelon and homemade ice cream or snow cones. Jose kept a cot there and would relax, nap or read the newspaper when he came home from work in the summer months.

The elementary school was a beautiful building, built in 1921 from concrete blocks used widely in construction at the time because they kept buildings cooler in the summer and warmer in the winter. The school consisted of two large classrooms, an auditorium with a stage and dressing rooms and a basement be low the stage. The basement was used as a kitchen and cafeteria during school and at PTA meetings. This school was demolished in 1950. The present day (2001) Hinkley School is on the same site on the corner of Hinkley Road and Santa Fe Drive.

Arredondo Children

All the Arredondo children attended Hinkley Elementary with the exception of Elvira and Jesse. Some of them attended for just a few years and the three youngest, Annie, Ramon and Ruben, attended the full eight years. The eldest children studied hard and led the way for the younger ones to follow, so they also became good students and good athletes.

Elvira (Vera)

Elvira resembled her grandmother, Paula, in her facial features and physical frame; she was the tallest of the five daughters and was very fair in complexion and hair coloring. When the family lived in Amboy, Elvira attended high school in Phoenix where she lived with her Aunt Alejandrina and Uncle Jesus Godinez. When the Arredondos moved to Hinkley, Elvira attended her last year at Barstow High, graduating with good grades. Jose and Herminia were quite strict with all the children but Elvira particularly caught the brunt of strict discipline upbringing—as the first child. Jose was so protective of his eldest daughter he was reluctant to let her marry without a long engagement to Antonio Salgado. Vera was chaperoned on all her dates by at least one family member sometimes two, even three or more. Tony joked with his brother-in-laws, that Vera's family had enjoyed "their" dates and that he had gotten to know her sisters just as well as Elvira when they were finally married. When Tony asked for Elvira's hand in a private meeting with Jose at the Arredondo home, Herminia later said Jose was biting his tongue feverishly –which was always an indication of Jose's anger, stress and jealousy—all worked into one big emotion. Tony also often joked that he was perspiring quite readily especially when he noticed Jose's Winchester 30-30 leaning behind the door of the living room. (We really don't know for sure if the part of the 30-30 rifle was true or not, but it is possible that Jose had the rifle there to show Tony how serious the meeting was to him.)

Antonio had a pleasant personality and was popular with the Mexican population in Barstow. He became close to the family, like a son and big brother, and the family was quite fond of him. Tony was good and generous to the family and Herminia thought the world of her son-in-law. He would take Ramon and Ruben with him, Elvira and their daughter, Eleanor (Lolly) to baseball games at Gilmore Stadium in Hollywood and the old Wrigley Field, home of the original Los Angeles Angels in the 40s. Antonio Salgado was from the State of Guanajuanto, Mexico, and also worked for the Santa Fe at the Casa del Desierto Railroad Station in Barstow. Tony saw many movie stars travel through Barstow in the 30s and 40s.

While Tony had an out going nature, Vera was quiet and reserved. Those who knew Vera well, however, knew she was witty and was a good storyteller. Her niece Jackie Arredondo summed it up this way: " Aunt Vera was a lot of fun to be around and when she told us stories, she would tell them with a straight face—like Jack Benny—which made the stories even funnier!" The Arredondos have always enjoyed sitting around telling childhood stories and reminiscing during and/or after dinner. Vera was a very good cook using her mother's principles to use as many fresh vegetables in her meals as possible and buying fruits and vegetables in season. Vera made the best peach cobbler from fresh peaches picked from her trees in their backyard. Vera would only serve her company the freshest baked good and would never have served anything not baked that day; the day before would be too old! Elvira was always extremely close to her parents and Aunt Lupe, visiting them almost every day. She demonstrated a deep love and devotion to her parents that is uncommon today.

Vera and Tony had one daughter whom they loved dearly. Paula Eleanor Salgado attended Barstow High School, was active in many clubs, lifetime member of the California Scholastic Federation and graduated with honors. Eleanor went to the University of Southern California (USC) where she started dating her future husband, Chris Connor, who was also a student there. Chris' parents were long time residents of Hinkley. Both Eleanor and Chris went on to get their Masters Degree from UCLA. Eleanor also has a Ph.D. in Spanish from Columbia University where she is presently a professor. Chris recently retired from AT&T as one of the vice presidents and is a private consultant in communications. They live in Millington, New Jersey and raised three children who are well educated and have successful careers.

Paula

Paula, the third eldest, was very studious and studied three to four hours every day—even on weekends. Her hard work paid off because she was valedictorian of her graduating class from Barstow High School. In previous years the valedictorian had also received a cash scholarship but instead Paula was given a trophy. Perhaps there was some discrimination…Paula was ambitious, worked in high school to save money for college. She attended the prestigious Woodbury College of Business, located on Wilshire Boulevard in Los Angeles and graduated cum laude and also received the Phi Gamma Kappa Bronze and Silver Awards. Her determination in attending college was ambitious since not many women did in the late 1930s and early 40s especially if their family wasn't well to do.

Upon college gradation Paula went to work at Camp Pond in San Bernardino as a senior secretary. Again, Jose's Spanish culture of protectiveness was displayed when he sent Aunt Lupe to live with Paula as her chaperone and housekeeper. Later, Paula's younger sisters, Minnie and Annie, lived with her and Aunt Lupe while they attended Ferguson Beauty College. Fela also lived there when she was working at Camp Pond. Aunt Lupe, who never married, loved her nieces dearly and was truly their second mother. Aunt Lupe said she couldn't speak or understand English but, of course, she did but not when it wasn't convenient! During the war (WW II) when there was food rationing stamps, Aunt Lupe delighted in confusing the butcher getting more than her meager allotment.

Paula met and married Elwood Grundei, a Master Sergeant, stationed at Camp Pond during the war, who was from Minnesota. Later when Elwood was transferred to Camp Lejune, North Carolina, Paula was able to transfer to another job within the Civil Service system. They were also transferred to an army base near Miami, Florida. After Elwood's army discharge they moved to Barstow where he started working for the Santa Fe in the refrigeration department. Paula stayed at home to care for their small daughter, Linda. Paula had planned to work when Linda started first grade, therefore, did not send her daughter to kindergarten giving them an extra year together. It was uncommon at that time for married women to have careers outside the home thus "Aunt Wally" as the Arredondo children called Lupe went to live with the Grundei family to care for Linda.

Aunt Wally had never married, almost always living with Jose and Herminia. She was a godsend to all the family helping when they were sick and became a second mother to Linda. When Paula returned to work it was at the Marine Base for Major Watson, a high profile officer on base. He had many contacts with the community and at one meeting at his office; he introduced Chief of Police, Floyd Howard, to Paula. This was about 1953 and Howard was only the second police chief for Barstow (previously they had a Sheriff). Paula must have made a big impression because within a couple of days, the Chief offered Paula the position of Police Matron and Secretary. Although Paula liked her job at the Base and leaving would be a cut in salary, loss of a nice office and filing clerk, Paula accepted his offer. She saw the challenge in being the first much needed police matron and secretary and she definitely liked the hours and commute better.

Barstow Police Department

When Paula started her job, the police department was located above the Fire Department on First Street by the bridge. There were steep stairs to the second floor where a couple offices housed the Chief and his staff. The jail was located down the street. Across from the Chief's office was the small City courtroom and Judge Powell's office. Since Paula was fluent in Spanish and had shorthand skills in Spanish, she also became the official Spanish court interpreter.

Not long after, the police department was relocated a few blocks west on Cozy Lane behind the Katz Restaurant on Main Street. The small building facing the alley was used as police headquarters where suspects and prisoners were questioned and/or booked on charges. Adjacent was a small jail with individual cells. Usually the cells were for holding rather than detention. Inmates with long-term sentences (7 seven days or more?) were transported to San Bernardino or Glen Helen. There was a two-bedroom house behind the jail that was utilized as offices. The living room was Paula's office with most of the file cabinets, the large bedroom and bathroom, was Chief Howard's office (later Chief Pierces), the smaller bedroom and bathroom was Captain McIntyre's office. Upon McIntyre's retirement, Sonny Davis became Captain and eventually Chief after Max Pierce passed away. The kitchen was used as the fingerprinting lab.

Special Projects

One of the first things Paula did after being hired by the PD was to organize all the paperwork. The record keeping and reports that had been shared by the officers had no order, and there were duplicate and/or incomplete records. Paula reorganized and set up the files on the Dewey Filing System and standardized their reporting system to Sacramento and the FBI. She initiated the procedure of typing up the name, offense, and details of each individual booked on a 3x5 card and was a cross-reference to the police file containing their fingerprints. This was quite efficient in absence of a computer. At that time the licensing of bicycles was mandatory, however it was disruptive with people always coming in so Saturday mornings were designated for that purpose only.

Chief Howard incorporated Paula's knowledge and enthusiasm in many of his projects such as improving the Trustee Program and bidding of restaurants to provide meal service to the inmates. Trustees were those inmates who had slight offenses that performed duties instead of being locked up in jail during day. Paula observed they had no accountability, nor were they doing their job. Under the Matron's watch they started doing all of the cleaning of the jail, offices and grounds and were also assigned to wash the police cars as well as those of the staff. Paula utilized the trustees' skills for instance if they had auto mechanic skills they checked the cars and painters or carpenters fixed up the jail which was old and in bad repair. Restaurants had to start bidding for contracts to provide the prison meals. The Katz or Village usually won the bids in those years.

Chief Pierce

After Chief Howard passed away around 1957 or 1958, Max Pierce became Chief and he also depended upon Paula. He had the City provide for a pistol range and his officers had a minimum time they were required to practice. As Police Chief he joined various law enforcement organizations in an effort to modernize the department and make it more professional. In 1959 he authorized his matron to wear a police uniform consisting of a khaki and tan wool jacket, tan wool skirt, and khaki wool police cap. Under the jacket Paula wore a tailored white shirt with a black wool tie. She wore the same police tie clip, only smaller, that the men wore. Her police badge, which also was smaller, was worn on the pocket of her jacket. Paula's handbag was black leather that had compartments for her gun and handcuffs. Inside the bag was a utilitarian whistle on a silver chain as well as her small leather police ID with her photo on one side and a second badge on the other side. Chief Pierce required his officers to buy their uniforms and their .38 caliber revolver stating "a revolver was more reliable than an automatic."

Paula's job was demanding, time consuming and required her to attend meetings and other activities in the evening. Her duties alone as secretary were full. As Police Matron she processed the female prisoners, questioned, interrogated, searched and booked them. She accompanied the transporting officer to San Bernardino when there was a woman prisoner and also when Sergeant Fernando Montoya was unable to escort a Spanish-speaking prisoner. After the PD purchased cars with air conditioning (probably the first A/C cars in Barstow) Paula liked the all day trip, stopping at Mt. Vernon Street on the way back to buy Mexican bread for the family.

Although Paula was very busy and not in good health, she still made time her for civic duty. She was active in the early 1950s at the USO when there many army men and marines stationed in the area. She and other ladies would make cookies and other goodies to serve the military. They organized dances and special events at the USO. She was active in the Barstow Professional Women's Club as was her close friends Judge Ava Powell and [check 1st names] Barstow Mayor Ada Pleasant. The three of them attended various city and law enforcement meetings and functions and had been interviewed by the *San*

Bernardino Sun. At that time, Chief Pierce used to jokingly say, "Barstow was run by Petticoats!"

Family Life

Paula loved her family, enjoyed their company, liked to entertain and loved life. Paula had looked forward to working in the new police department on Mt. View. Tragically, she was born with a defective heart and passed away (in her early 40s) before its completion in 1962. Paula didn't see her daughter graduate, marry or get to know her grandson, Jim (who now lives in San Clemente with his small daughter) or get to know her younger nephews and nieces, Ramon's and Ruben's children. And sadly they didn't get to know their grandmother and aunt who was a warm and generous person. Paula was known for her compassion, charm and willingness to help all people. Many inmates later said she could be tough if needed but she always treated them with respect and dignity. Many years have passed since her death but she is still missed by her family, especially Linda, her daughter. Linda graduated from the University of Nevada, Las Vegas with a B.S. and is employed as a Social Service Supervisor for the State of Nevada Welfare Department in Las Vegas. She is married to George R. Parson, CPA, who has his own local accounting firm.

Jesse

Jesse from his early years was quiet and reserved with an unusual personality. Although he was quiet, he was lovable and affectionate in a different way. He had an inherent manner of respect and an abundance of patience with people and was loved and respected by all the people he came into contact with! Jesse was a natural born mechanic and his passion was taking electrical and mechanical apparatus or gadgets apart to see how they worked and how to put them back together. Jesse helped Butch Palmer in his Hinkley Garage and was constantly working on Model Ts and Model As. He did not inherit this talent from his father Jose. When gadgets needed repair around the house like replacing a doorknob or a light switch, it was Herminia or Jessie who fixed it. Jose loved to read and was current on world events and politics. Herminia was the practical one and the family disciplinarian and Jose, his children teased, was the "philosopher!"

Unlike Ramon and Ruben, Jessie was a very private person who avoided arguing, heated discussions, and political or formalized social activities. He liked to play his mandolin and belonged to a baseball team comprised of young Mexican men. Jessie played a solid center field for his team. Some of his friends on the team later became very good athletes at Barstow High like Chevo Cedillo and Joe Renteria, Sr. Chevo was a great baseball player and boxer; while in the Army reportedly sparred with the great Heavyweight World Champion, Joe Louis.

Jesse was a big help to the family, always minded his parents, but did not like school. Instead of finishing high school, he wanted to go to automotive school and become a mechanic. When Jesse was 16, he attended the National Technical Automotive School on Figueroa Blvd. in Los Angeles, graduating when he was 18. Jesse loved the school and his instructor. He had finally fulfilled part of his dreams and couldn't wait to get started on his goals to be a great mechanic and own his own garage.

Jessie became ill a few months after graduation and died of pneumonia the age of 18 in 1932. His sisters Elvira, Paula, Minnie and Fela felt that they lost not only a brother but also a loyal and lovable protector. He once fought a big bully who was harassing Elvira at school. After Jesse beat him up (after several warnings), the bully never bothered Elvira again. Jose and Herminia agreed that his passing was the most tragic time of their lives, perhaps because he was the eldest son and so young and because of his quiet and lovable demeanor. This was a devastating time for the Arredondo family as Jose's brother Miquel and his mother Paula also passed away that same year.

Herminia Gloria (Minnie)

Minnie played basketball and ran track in school. She was quite the sprinter in track and placed first in her grade level in a number of track meets in the Barstow-Victorville area. Minnie was a good all—around student, but loved sports and was as competitive as any one of the Arredondo children. Minnie would probably have done very well both in scholastic and athletic activities had she gone on to college. After graduation from Barstow High School, Minnie took various jobs in Hinkley and Barstow. By the time World War II broke out, she was living in San Bernardino attending Beauty School. During the war, she also worked for the U.S. Army at several installations around San Bernardino and also was a volunteer for the U.S.O. on Arrowhead Ave. Minnie participated in a number of war bond sales drives, volunteering many hours in the war effort.

Minnie was a thoughtful person, especially with her younger brothers, Ramon and Ruben. When Ruben was 14 or 15 she found out that Ruben wanted an archery set for Christmas (during the war and many items were either difficult to find or not available) but she spent many hours trying to find it. Finally she found a nice set that was beyond the quality that Ruben ever dreamed of receiving for Christmas. "I must of used that archery set for ten years and as I got older I appreciated it even more—my sister Minnie made that Christmas for me!

After her sisters married and left their house in San Bernardino, Minnie and Aunt Lupe returned to Hinkley. In 1947 Minnie married William "Bud" Carr who was working for the Santa Fe in Barstow. Bud a native of Oklahoma had been in the service, had been in active duty in Europe and upon his discharge settled in Barstow. Their only child, Buddy Joe, was born while they were living there. In the early fifties, they moved to San Bernardino where they resided for several years. Bud switched careers leaving Santa Fe to work for the Federal Civil Service System and in 1962 was transferred to the Sacramento area.

Minnie inherited Jose's love of politics and she always was up to date on news. She did not inherit her Dad's singing voice as hers was lovely. She would lead her sisters in singing Christmas carols and Mexican favorites while Paula and/or Eleanor played the piano at holidays. Minnie was an active member of the Democratic Party in the Roseville-Orangevale areas (near Sacramento). She also spent time registering people for the party and participated in other party activities. Minnie loved to write and had a very good memory; she wrote many beautiful letters to family members. Minnie was a lovable person who one could love easily as she spread her affection and charm to all that knew her. Minnie passed away in 1997 in Orangevale and is survived by her husband and their son, Buddy Joe, his wife Sharon, their two sons, daughter and her three small children. Minnie was a wonderful grandmother to her three grandchildren and was always taking them out to lunch or dinner and loved to give them gifts. Her grandchildren miss their grandmother immensely. Buddy became active in sports while in high school and excelled in softball and golf. Later he became a very good amateur golfer (takes after his dad) in the Sacramento area. Buddy has many years of retail management experience and currently has his own large K-Mart Store in Sacramento by Arco Basketball Arena.

Rafaela (Fela)

Fela was active in several clubs in Barstow High School, was a honor student and participated in sports. Fela made the California Scholastic Federation (CSF) rolls several semesters. The CSF honored students who maintained the highest grade point average level for several semesters. While attending high school, Fela helped the proprietor of the Hinkley Store and Post Office. Fela worked in the Post Office assisting with mail and also helped in the store waiting on customers and stocking canned and dry goods. Fela continued to work for Mrs. Burns about two years after graduation.

At the beginning of the war, Fela along with Minnie moved to San Bernardino and also worked in several Army posts in the San Bernardino/Riverside area. Fela was also active in volunteering for the U.S.O. program in San Bernardino. In 1944, Fela met Robert (Bob) Schultz at a dance in San Bernardino. They liked each other and started dating regularly and in December of 1944 they were married in Las Vegas, Nevada. Bob was from Wisconsin but, like his other brother-in-laws, did not return home and instead stayed in California after being discharged. Bob went to San Bernardino College on the G.I. Bill and majored in Business. After graduation, he went to work at Mobile Oil Company in management.

Fela, being a Virgo, represents that sign well and was the most organized person in the Arredondo family. Fela was and still is great at organizing events and is a stickler on details. Fela was a wonderful homemaker, especially when her two children were in school; her home always smelled like fresh baked pies or rolls. She is a meticulous and good cook also (Fela was cooking "healthy cuisine" in the early 1950s). Fela stayed at home with their two small children who were close in age. Although they lived in San Bernardino, they always made it to Barstow during the holidays. At Christmas Fela arrived with beautifully wrapped presents and delicious homemade cookies and pies. Fela, the Virgo, can be picky (like Ruben, also a Virgo), but she is a devoted and loyal person to her immediate family and her brother and sisters and their families. Her actions truly speak louder than words. Bob and Fela had a truly great relationship until Bob's death in 1995. Fela, and her sisters, worked hard at being excellent wives.

Fela today lives alone in the San Fernando Valley. Her daughter, Barbara, and son, Robert, also live in the valley, not too far from their mother. Barbara is a registered nurse, has a Bachelors and Masters Degree in Public Health and loves her job as a nurse for the Los Angeles School District. In the summers she consults as a Nurse Practitioner. Robert was the first of the Arredondo grandchildren to retire after a career in beverage sales. Now he has more time to track his stocks as a day trader.

Mary Ann (Annie)

Annie excelled in track and basketball but her top sport was track and field because of her running ability. She ran the sprints and won first place in a number of track meets in the San Bernardino and high desert areas. Annie became the top girl sprinter at Barstow High for two years. Annie was popular with her teachers and classmates because of her friendly personality. She was an Honor Society student and belonged to several clubs where she worked to advance student concerns and problems in student government. Annie received the Citizenship Trophy when she graduated. This trophy was presented to the graduating student who exemplified outstanding abilities in sports, scholastic and student government activities.

While in high school Annie began to date George Vallis Jr. George lived in Minneola with his parents, two brothers and two sisters. George Vallis Sr. was the section foreman of the Minneola Station. George Sr. was of Greek extraction, being born in Greece, He was well known in the Barstow, Newberry and Minneola areas, and the "trackeros" called him "El Griego" or the Greek. His wife Chonita" was from the State of Michuacan in Mexico, hence Vallis could speak fluent Spanish as well as Greek and English. It must be noted that Vallis served on the Minneola School District Trustee from 1937 thro 1940, as noted in *A Piece of Baling Wire: Homesteading in the Desert, Appendix B,* by Grace Loomis Odell.

George Jr. played football for Barstow Hi in 1939 and 1940 and was one of the top lineman of the era. He played guard and tackle and was a member of the undefeated squad of 1940, under the outstanding coach Pete Clentzos. George went on to San Bernardino Valley College where he made the All Eastern Conference teams two straight years. He was offered a football scholarship at USC under Jeff Corvath. However the war came.

George served with the 8th Air Force as a radioman on a B-17 and flew 17 missions.

Annie and George Jr. married in 1946 and made their home in the Bay Area.

George and Annie have two sons Ralph (named after his great grandfather Contreras) and George III. Ralph and Georgie (as the family affectionately calls him) played football in high school and in junior college. Ralph went on to play at Cal State University, Hayward, where he was a star linebacker. Ralph graduated with a teaching degree and today teaches and coaches at his alma mater high school, Mt. Diablo in Concord. He married his high school sweetheart, Connie, and they have one son, Daman. Georgie is a top-notch manager of a Safeway store in the bay area and when he was promoted to manager of his own store, he was the youngest man to make manager in their chain. He and his wife, Mary Francis, have two daughters, Elyse and Ashley. Throughout the years Annie and George have gone to all the sporting events of their sons and to those of their grandchildren.

Ramon

Ramon and Ruben (Benny) were average students but played football, basketball, track and baseball and lettered in a couple of these sports in high school. When Ramon was in high school many athletic events were suspended or cut back because of the war; however, he was able to participate in football and wrestling, becoming a good wrestler. Ramon made the school's first team (football) as a blocking back in his junior year in the old single wing formation. Although he only weighed 130 pounds, his blocking was superb gaining him the respect of his coaches and teammates. Ramon, because of his size and will to be involved in blocking for the running back, got hurt quite often. Ruben remembers him getting the wind knocked out of him in many games, having to sit out a few plays to regain his breath, then always returning to the playing field. Ramon told his brother several times that he loved to level the linebackers, especially when they were not expecting to be blocked by the "little short blocking back"! Ramon (like Annie) was popular in school and in college and always had many friends. Ramon was the most outgoing one in the family and knew everybody. His mother used to say, "hasta los perros lo conosan" (even the dogs know him)!

In his senior high school year, Ramon was drafted into the Army at age18. Ramon knew he would be drafted so the summer before he had tried to enlist in the Army Paratroopers but was not accepted because of his curvature of the spine. Ramon took basic training at Camp Roberts near Paso Robles. He showed leadership qualities early in his training and became an acting Corporal of his training platoon. In March of 1945, Ramon was sent overseas as an Automatic Rifleman with the 32nd (Red Arrow) Infantry Division in Luzon, Philippines. He saw action for six months, distinguishing himself as a mature young soldier. His platoon leader sent his parents a nice letter commenting how well Ramon performed in combat conditions. Ramon earned the Infantry Combat Badge and his unit, while in combat, received a Presidential Citation. Ramon was promoted to a three-stripe sergeant at age eighteen.

Eighteen-year-old Sgt. Ramon Arredondo in 1945 while in the Phillipines.

After the Japanese surrendered, Ramon's division was one of the first units to occupy Japan. Ramon finished high school after WW II was over then went on to Chaffey College for three semesters and graduated from Orange Coast College with an A.A degree in agriculture. He worked for a short time as an agricultural inspector in Truckee located in northern California.

One incident stands out that would reflect Ramon's personality and his stand on civil rights: In 1947, Ramon and his parents were looking for a house to purchase on his G.I. bill on the west side of Barstow. They found a home on Nancy Street and a sale agreement was drawn up with the paper work and financing to be done by the Bank of America. Shortly after the sale, Ramon was notified that the bank would not finance the sale because homes in that area could not be sold to Mexicans. When Ramon got the word, he became furious and arrived at the local B of A, cussing out the manager, the loan officer and anyone else in sight! He told the assistant manager that he "did not sit in a filthy fox hold fighting for a son- of-a-bitch like him that was going to tell him where he could buy a home." He threatened to call the Veterans Administration and turn the matter over to Congressman Shepard's office. The home sale was finalized in about thirty days after the incident.

Ramon went to Mexico in 1957 to help his father work the ranch in Irapuato for ten years. Ramon's college and experience was an asset in growing wheat and corn and the envy of the other ranchers. Ramon married a young pretty girl named Esther in Irapuato and they had three boys, Alex, Oscar, Ramon Jr. and one daughter, Paula, before selling the ranch. Ramon moved back to the States (Barstow) with his family and returned to Hardwicks where he had worked before leaving for Mexico.

A few years later Ramon went into car sales and was called the "Little Man with a Big Heart." Ramon was very family oriented and had his children in all kinds of sports. More than once he worked two jobs to give his family as much as possible. He was proud, as was all the family, when he oldest son, Alejandro, graduated from Optometrist College and later opened his own office in Huntington Beach. He was also proud and fond of Paula's two little boys who lived around the corner from him. Paula works for Stater Brothers grocery and is married to Ernie Langford who is employed at the Marine Base. Ramon delighted in his sons Oscar and Ramon, Jr. Oscar who lives in Brea with

Alex is a salesman for Pepsi Cola. Oscar would take his father to the fights in Las Vegas to see Julio Cesar Chavez and Oscar de la Hoya. Ramon Jr. lives in Barstow and is also employed by Stater Bros. The family recently lost Ramon (Oct. 2000) who fought gallantly against cancer. To the end he had a sense of humor and loved to tell and hear jokes. We miss you, Ramon.

Ruben (Benny)

Ruben, while not a great student, liked school probably because he always had two or three good friends who, like him, loved to play football and baseball whenever they could. Ruben had a good friend, Buddy Larson, with whom he remained very close with throughout their school years at Hinkley and at Barstow High. Ruben and Bud learned a good passing combination in their two years of varsity football (Benny threw to Buddy). Benny in his senior year was the starting quarter back of his team that won the Desert Inyo League but then lost to Chino High in the CIF semifinals. Academically, Benny had three teachers at Hinkley Elementary who had a lifetime influence on him: Miss Blevins, Wilber Olsen and Phil Capen. Mr. 'Capen not only influenced Benny to be tolerant with all people of different backgrounds and ethnic origins but Ruben and others considered him an exceptional teacher. Capen would not tolerate any race prejudice or bigotry in his classroom or in the entire Hinkley School.

He once became enraged and chewed out a girl after she told a short poem denouncing Jews, Catholics and black people—of course the word black was not used. Mr. Capen, however angry, handled the situation very well and gave the whole classroom a lecture on how all of us were Americans and that we all should practice tolerance towards all people off different races and religious backgrounds.

After graduating from Barstow High, Benny worked for the Santa Fe for one year, then went on to pursue his education at Orange Coast in Costa Mesa where he majored in Business Administration and Labor Relations. He lettered in football and enjoyed the students and teachers of the beach city community. He still communicates with several friends that live in the Newport-Costa Mesa area. Benny believes that his studies and work in labor relations helped him years later when he served as a board member for his labor union (as an PG&E employee) representing workers from the bay area to Needles.

In 1953 Benny entered the Army after serving two years in the Active Reserve and was stationed at Ft. Benning, Georgia, where he played football for the Special Troop Command. He met and played with many big-time players in the college and pro ranks at that time. He became a good friend of his coach, Andy Kozar, the famous fullback from the University of Tennessee, whom he kept in touch with for several years. After two-years active duty, Ruben returned to his job at PG&E retiring in 1989 with 36 years of employment.

Benny believes in serving his community, has served on the Barstow Park and Recreation Board of Directors and is currently serving as a Board Member for the Barstow Fire Protection District. Benny has served on the Fire District for many years and is proud of the new fire station on West Main Street that was constructed during his tenure as the Board Chairman of the Fire District from 1998 to 2000. Ruben is also active in the Mojave Democratic Club. He served two terms as club chairman and worked for several candidates, such as Senator Ruben Ayala and Barstow's own Congressman Joe Baca.

Ruben likes to think that he has been guided through life by simple terms taught him by his father Jose. One is: Remember you are just as good as anyone—but at the same time you are no better either. How you behave throughout life is what's really important! Another is: You don't have to like people, but you do have to always show them respect!

Ruben has five daughters, the oldest three; Annette, Claudette and Jackie are from a previous marriage. He has been married to Elizabeth Caughlin Lilly for 27 years and they have one daughter, Jessica. Liz has a daughter from a previous marriage, Kimberly. Ruben met Liz in 1972 and she helped him through a tough time—at that time his lifestyle and life had changed dramatically after his divorce in 1971. Liz helped with raising Ruben's three daughters, plus her own Kimberly and Jessica. Benny is very grateful for the great job she did and considers Kim to be like his own daughter, an Arredondo.

Annette lives with her husband, Kirk Jefferson, in Chico. Kirk is a long- time employee of PG&E. Annette has a Master's Degree from University of California, Berkley and presently teaches. They have a son and daughter. His second daughter Claudette was Miss Barstow in her senior year and also second runner up to Miss California that year. Claudette has a B.A. degree from Cal State University, Stanislaus. Claudette is married to Philip Dale. They reside in Corona with their two small sons. Jackie, Benny's third daughter, is the only one living in Barstow. She works at Smart & Final and keeps busy with her daughter, Taylor.

Kimberly is the owner of the Mermaid Beauty Salon and Day Spa in San Diego with sixteen employees. The youngest daughter, Jessica, lives in Las Vegas and is a junior at the University of Nevada, Las Vegas, and plans to major in Communications. Jessica and Kimberly are not married as of 2001.

Leaving Hinkley

In 1953 Jose Arredondo retired from the Santa Fe Railroad with close to 50 years of service. During that time, neither Jose nor any of his crews had ever had any lost time from an injury or accident. This must have been some kind of record; however the family is unaware of any acknowledgment, neither did the Santa Fe present Jose with a watch or momento of his retirement. His Division Engineer did send Jose a letter thanking him for his many years of dedicated service and he also received a congratulatory letter from the First District County Supervisor Cunningham. Family members have saved these letters. However, the congratulatory letter sent by Congressman Harry Shepard was lost.

Jose did not expect to be thanked or acknowledged as he was a model company man for the Santa Fe, dedicated and believed an employee owed all his loyalty to his employer. The railroad had recognized Jose's leadership skills and work ethic early on using him to trouble shoot as section foreman at the different stations transferring him many times. His work ethic was such that he didn't miss work for sickness or take shortcuts—he simply did his best. This was a value he passed on to his family.

Jose's retirement meant they would no longer be living in their house in Hinkley and that they would be moving to Barstow. They would be leaving an era of good friends such as the Jacksons, Palmers, Dolphs, Larsons, Duartes, Parkers, Hernandez, Harris, Hubbards, Hartwicks and others. Some of these families had also moved, but at one time they had gotten together at school events, sport activities, birthdays and holidays. During the summer and after school the families had a great time playing ball in an area cleared between the foreman's house and the Hinkley Garage. Everyone played including the mothers and fathers. Herminia or Mrs. Larson would usually make delicious homemade ice cream while their husbands threw horseshoes.

The older Arredondo grandchildren loved the house in Hinkley; to them it was a paradise. The house sat on two acres surrounded by a natural fence of tamarisk bushes with a big drive in entry. In the yard there was a high rise so large that grand kids pretended it was their "hill" riding down on bikes and wagons. The fruit trees, big garden, chicken coop and cellar were entertaining novelties to them. Also in the yard facing the tracks was a raised platform of gravel with large white rocks spelling "HINKLEY" to identify the depot and station train passengers. Many times the kids would stand on it and wave to the passenger trains; people always smiled and waved back.

Remembering the War Years

In 1944 train travel increased 100 percent. It was not uncommon for 30 or more troop trains to whiz by Hinkley going 70 to 80 mph for embarkation in Oakland and San Francisco. During those war years, Jose would protect the Santa Fe property with a passion even more so than if his own property. If there was a derailment or materials fell off the flat cars, he and his crew would take extra care to secure these. He would threaten anyone found tampering with the goods by calling the constable or sheriff. On his own initiative, he often made night patrols to the derailment areas using an old fashioned motorcar or by being driven to the site by Ramon or Annie.

The day Pearl Harbor was attacked an aircraft warning system was set up at the Hinkley Depot. The system's job was to spot and identify all aircraft flying through the area by phoning a central headquarters station either in Barstow or San Bernardino. The Depot was initially used because of its telephone and telegraph facilities. Later a room was built next to the elementary school that had windows all around so that the aircraft spotters could keep track of the sky during cold weather. Everyone in Hinkley participated in this warning service working in shifts of two to four hours. Annie, Ramon, Fela and Minnie all contributed time to the effort especially Ramon and Annie who worked many hours after school.

During the War, Congressman Harry R. Shepard was once a guest of the Hinkley School PTA where he and Jose met and talked. The Congressman and Jose shared a strong Democratic Party philosophy and the two became good friends. The Democratic Congressman stopped by at the Arredondo home whenever he was in the area. He knew that Jose and Hermina were not U.S. citizens, as they needed to retain their Mexican citizenship to keep the land they owned in Mexico. Harry would still leave his cards and reading materials with Jose to pass out to the people of Hinkley, Hawes and Harper Lake. He told Jose, "You will help me get elected even if you only get a few votes for me." Jose worked hard to get him more than a few votes! Congressman Shepard, during and after the war, was probably among the top three most powerful Congressmen in the House of Representatives.

Ruben well remembers the Congressman as a down to earth kind of old-fashioned politician always smiling and shaking hands enthusiastically with everyone even the children. Whenever Jose heard the National Anthem or Flag Salute over radio broadcast, he would stand and have his family stand as well. He admired President Roosevelt and Winston Churchill and followed them in the news as well as reading books about them in the years to follow. Linda, Jose's granddaughter, thinks if her grandfather had had the education and opportunity, he would have been a good Statesman.

Jose was not an educated man but he was a wise man. He did the best he could for his family by encouraging them to further their education beyond high school.

Jose And Herminia After Retirement

After Jose retired, he, Herminia and Aunt Lupe moved to their house on Nancy Street that had been purchased a few years before. Not that long after, they moved into their lovely new house they had built on First Street. Jose, now in his 70s, grew restless wanting to go to Irapuato to live and work on his ranch, La Soledad. Herminia had other plans and told Jose she, too, had worked very hard living in hot, desolate areas for many years. Herminia knew she would not have the same comforts on the ranch. She liked her new house, the location and would miss her children. She saw Vera and Paula on a daily basis and Ramon and Benny still lived at home. Fela and Minnie visited often from San Bernardino. Herminia was an admirable woman and good role model to her daughters and granddaughters. Nutrition back then wasn't an issue, however, Herminia used as many fresh vegetables and fruits in season as possible. Her snacks in the evening were walnuts (which she cracked) mixed with raisins making her own trail mix. Herminia passed on so many good traits that it is no wonder the family loved and respected her so much. Yes, Herminia remained in Barstow and continued to enjoy her family and they also continued to spend the holidays with her and Aunt Lupe. That is not to say that Jose and Herminia did not visit one another. Their trips to see each other were not as frequent as they have liked but usually the stays were a month or so in duration—quite the modern marriage!

Jose left for Mexico where he still had cousins, nephews and nieces. He rode horseback overseeing his ranch checking crops, equipment and animals putting in many long hours of hard work. Ramon wanting to help his father on the ranch moved to Mexico. After living there a couple years Ramon thought there might soon be a drought and made some type of water reservoir. He had warned his neighbors in vain to do the same. Sure enough the following year was an extremely dry spell. One of their ranch hands warned Jose that some jealous and angry ranchers were going to destroy the reservoir at La Soledad. When these ranchers got there, he and Ramon were actually waiting for them with a shotgun and a pistol. They quickly backed off; Jose and Ramon had quite a reputation after that! Lupe also moved to the ranch

The Arredondo family in Barstow in 1985.

for a couple of years to cook and keep house. Ramon told the story of Aunt Lupe getting sick and rushing her to a doctor over rough, bumpy washed out roads. Aunt Lupe told him afterwards to let her just die the next time she got sick! Jose worked his ranch until he was about 87 years old. The ranch with its house, school and chapel was sold but the new owner kept Jose's foreman and other ranch hands.

Extended Family Reunion

In 1995 the Arredondos (distant cousins) living in Mexico hosted a BIG family reunion in Irapuato inviting their U.S. relatives as well as relatives in Spain. Ramon, Esther, Alex, Oscar and Ruben knew some of the relatives there but Robert, Linda and her husband, George didn't know any of them but they all went for the three night festivities. There were over 150 Arredondos in attendance. The highlight of the trip was when Ramon took his family (from California) to the La Soledad. His old foreman who still worked there was delighted to see Ramon again and had his wife cook a delicious lunch for them consisting of fresh turkey mole, warm homemade tortillas, chili, beans and rice. Ramon drove his family around the ranch to the see the canal, their old house, and the chapel. The last stop on the ranch was the little schoolhouse, named after and dedicated to Jose Arredondo and in the schoolyard stood a moving memorial of Jose. Many relatives in the U.S. and Mexico have become close, stay in touch and plan another big reunion in the future; Jose would be pleased.

Barstow Is "Home"

The Jose Arredondo family today is dispersed all over California, Nevada, New Jersey, New York and Florida, yet to most, Barstow is still home. Jose passed away in 1969 in his late eighties and Herminia in 1973 in her late seventies. However, many family members still go to Barstow (Uncle Ruben's) at Christmas, other holidays and on special occasions. Barstow is where they can pay tribute to their deceased love ones by taking flowers to the cemetery. The Arredondo grandchildren are closer than most cousins and still like to get together as they did in their childhood. Jose and Herminia, who were loving parents and grandparents, passed their family values and appreciation of family down through the five generations—this is their legacy.

The Arredondo family thanks the Mojave River Valley Museum for the opportunity of sharing our family history and giving insight to the important role the Mexicans played during the heydays of the railroads in the Southwest. This project has inspired us to continue our story focusing on the grandchildren and great grandchildren of Jose and Herminia. The Arredondos have been enriched by the different ethnic backgrounds of their children and grandchildren's spouses who have introduced their own culture and traditions. Although these cultures and traditions may have been quite different, they have blended well and have been enhanced.

Special recognition must be given to Ruben's wife Liz for all the help she gave him in writing much of this story. Liz helped with organizing the first dozen pages and witn doing the typing. Liz, a creative person, was a great influence to Ruben to go ahead with continuing the story for several years when Ramon and Ruben originally thought of the idea.

Ruben Arredondo's daughter Claudette Dale, Miss Barstow for 1979-80.

Elizabeth earned an A.A. degree in nursing from Barstow College and operates a Tiffany Lamp business with sales in the San Diego and Barstow areas.

The second recognition must also be given to Ruben's brother Ramon, Sr. Ramon first encouraged Ruben to start putting the early part of the story together about Jose and Hermenia and much of the information about their living in Mexico from 1910 to 1915 was provided by Ramon, Sr. Rafaela Schultz also provided much of the information of the early living in Amboy, Hinkley and Barstow.

Special thanks go to Queta Hernandez, Maria Mora Hernandez, David Villafana for their contributions to the family history and desert history. This story was compiled and edited by Ruben "Benny" Arredondo and his niece Linda Grundei Parson (Paula's daughter), who organized the family's story to make it a more concise version of the original.

Paula Eleanor Salgado Conner and daughter Paula Conner.

EPILOGUE

by Clifford J. Walker

The little babies brought from New Mexico, Arizona or Mexico or babies born in the early years of Barstow—now have retired and a couple even passed away. When they were growing up, all met some kind of prejudice, such as the Mexican family who always took his family to Newberry Springs to swim because some pools were closed to Hispanics on weekends. "Even after all the pools were integrated, my father still took us out to Newberry to swim—force of habit, I guess," said Ray Perea, Barstow College dean.

Now in Barstow any family can buy a house anywhere it can afford. The Bank of America that refused to sell two Foglesong division houses to the Arredondo and Yslas families would not dare turn down a loan and say "We can't sell houses there to Mexicans." Two families helped integrate Barstow by standing up for their rights as Americans and standing up for justice, and as a result have given more rights to all America and added to the thrust of equal justice and civil rights that pushed ahead in the 1960s and 70s. They set a precedent, which enabled black teacher Toy Owens to buy a house in the Montara area after being turned down because of his race. Barstow became truly integrated.

These children attended fine integrated schools with dedicated teachers who liked the desert, liked the towns and the students. All students who wanted an education—and some who didn't—received a fine education. Looking through the old yearbooks of Barstow High School and Kennedy High, one notices that the books are filled with students of these Latino families. They played sports, were as active as any group of kids, studied as hard, some excelling, some happy to get by with a C grade—typical kids, perhaps a bit more into school politics as many were elected student body officers. Normal kids: Meeting their first sweethearts, dating, dancing, breaking up, cruising around, hurrying to Victorville for the Friday night big game, either playing, marching or cheering. Practically everyone graduated. During World War II Ramon Arredondo, for example, tried to enlist in the paratroopers, but was turned down because of curvature of the spine. In 1944, however, he was drafted in the army in his senior year, served in the Philippines, and when he came back from the war, he earned his high school diploma.

The younger generation settled on their lifetime mates—sometimes for a lifetime and sometimes not. After high school the younger generation went to college while some stayed in the desert working at Santa Fe, or military bases such as the Marine Base and Fort Irwin.

Having grown up with thrifty parents who knew how to survive the depression and World War II, and by working hard, these new generations completed raising their families, and many of them are retired. .

Since these families have married into almost every ethnic group, they benefited from cultural diversity. In-laws expressed the love and acceptance they received from their new Hispanic relatives, and in the process all cultures have grown richer. Just as some families learned new foods and traditions when they moved to California from New Mexico or Arizona or Mexico, so too have the younger generations blended the Hispanic culture and have been enriched. Yet there are pieces of New Mexico left in the older family members. Even for those who live out of the Mojave Desert, part of the desert remains in them.

Although tortillas are in every supermarket in southern California and anyone can buy New Mexican chilies, the smell of fresh homemade tortillas cooking is still important to the next generation. So too, is the distinct aroma of chilies roasting in the back yard. Neighbor kids continue to come over to their Hispanic friends' homes and take in the unique smell of fresh-made tortillas, made the way a great grandparent made them. Spread with butter or perhaps some new spread the great grandparents never dreamed of, kids take their rolled-up tortillas outside and continue to play—together—in the streets of America.

I asked Ray Perea if his family used to sing "Las Mañanitas," and he answered, "Oh yes, and we still do." The next generation of children are awakened with beautiful words in Spanish like: "We come to salute you; the day that you were born all the flowers were born." Grey hair doesn't matter. Noberto Luna used to sing it romantically to his wife Annie in the kitchen when she was cooking dinner. He couldn't sing. "Ah, get out of here," she jokingly said pushing him away. In 2001, sisters, brothers, children, nieces and nephews sang it to Noberto's son Leo on his 62nd birthday. There was not a dry eye in the group, all thinking about their mom and dad and other ancestors. And another generation of children learn the tradition, feeling special, and feeling they belong to the family and the world.

Values of hard work, loyalty to their heritage, their community, and their families are still paramount, as is education. Most families developed community leaders, education, labor and church leaders. They are into every profession, are business owners, good loyal workers.

The Mojave River Valley Museum is proud to present this book and help complete a clearer picture of the history of the desert. What a tremendous experience, working with these five wonderful families. What a pleasure it is to see so many of my former students (who are not too mad because of the grade I gave them) and to see their families continue to grow and prosper and enrich the nation.

The future of these families looks great!

Grandchildren of Naydean Jaramillo Williams. Left, Andrew Sinner, top right, Josiah Williams, Celia Williams, Jessica Sinner and Dawson Hughes vacationing in Colorado in 1999.

Marissa Villareal and son Calob, granddaughter and great-grandson of Gillie and Vina Olguin.

APPENDIX A
JARAMILLO FAMILY

JARAMILLO FAMILY
ANCESTORS OF ANDRES AND PITA JARAMILLO

Loreto Jaramillo
b. Approx. 1858, San Juan, NM
Dionesa Chavez, wife
b. San Juan, NM

Nicanor Jaramillo
b. Jan. 10, 1878, Sabinal, NM
d. May 14, 1957, Albuquerque, NM
m. Encino, NM

Tomas Bachicha
b. Villanueva, NM
Emilana Baca, wife
b. Nov. 5. 1867, Villanueva, NM
d. June 30, 1937, Encino, NM

Maria Efrin Bachicha, wife of Nicanor
b. January 1, 1883
d. May 1937, Albuquerque, NM

Andres N. Jaramillo
b. May 20, 1919, Encino, NM
d. Sept. 19, 1999, Barstow, CA
m. Albuquerque, July 9, 1939

William Riley Milligan
b. Feb. 18, 1897, Reserve (Catron), NM
Rufina Chavez, wife of William Riley Milligan

Anna Maria Milligan, wife of Jose Antonio Gonzalez
b. June 22, 1871, Hillsboro (Sierra), NM
d. March 25, 1930, Las Cruces, NM
Jose Antonio Gonzalez, husband of Anna Maria
b. Jan. 1, 1837,
m. 1884, San Pedro, NM

Rufina Muerta Gonzales, wife of Tomas Silva
b. Nov. 2, 1891, San Pedro, NM

d. Oct. 11, 1984
Guadalupe "Pita" Silva
b. Jan. 18, 1922, Socorro, NM

Andres N. Jaramillo married Guadalupe "Pita" L. Silva on July 9, 1939, Albuquerque, NM.

The Jaramillo Family

Descendants Of Andrew And Pita Jaramillo

1. Andrew Jude Jaramillo
2. Naydean Viola Jaramillo (Williams)
3. Paulette Eunice Jararnillo (Purdy)
4. Geraldine Ann Delores Jaramillo (Simko)
5. Mary Ruth Jaramillo (Arguelles)
6. Lupe Elma Corine Jaramillo (Mauldin)
7. Anthony Joseph Jaramillo

1. Andrew Jude Jaramillo
Angie Hurtado, wife
Teresa Dean Jararnillo Townsend
Brian Townsend (husband)
Denny Townsend (child)
Timothy Townsend (child)
Bobby Townsend (child)
Andrew Jaramillo Jr.
Angelina Jaramillo De Rugoma
Jose Luis DeRugoma, husband
Tyler Jaramillo (child)
Nick Rugoma (child)

2. Naydean Jaramillo Williams
M.D. Williams, husband
David Williams
Susan Dyson, wife
Rindy Williams (child)
Isabel Williams (grandchild)
Isaiah Williams (grandchild)
Mike Williams (child)
Brittany Williams (child)

Andrew Williams
Jamie Day Williams, wife
Chelsea Saenz (child)
Amy Williams (child)
Josiah Williams (child)
Celia Williams (child)
Marion Williams Sinner
David Sinner, husband
Andrew Sinner
Jessica Sinner
Rindy Williams Hughes
Mike Hughes, husband
Dawson Hughes

3. Paulette Eunice Jararnillo Purdy
Thomas David Purdy, husband
Katrina Louise Purdy Hokkanen
Joel Hokkanen, husband
Krystina Paulette Fries
April Fries
Kathlina Marie Purdy Dillion
John Dillion, husband
Terina Thomasina Dillion
Genna Leigh Dillion
J. C. Dillion

4. Geraldine Ann Delores Jaramillo Simko
Edward J. Simko, husband
Ted P. Simko
Matthew Simko
Deborah A. Flores
Francisco Flores, husband
Edward D. Simko
Shawn Simko, wife
Lexie Simko
Kimberly L. Bas
Eric Bas, husband
Amber Bas
Eric Bas
Christopher Bas
Tyler Bas

5. Mary Ruth Jaramillo Arguelles
Alfredo Arguelles, husband
Paul A. Purdy & Michelle
Shoemaker, wife
Taylor Purdy
Dominic T. Purdy
Danielle A. Wilson
Alisha Wilson
Adrian A. Arguelles

6. Lupe Elma Corine Jaramillo Mauldin
Mark Mauldin, husband
Teri Lynn Mauldin
David Clutter, husband
Brent Clutter
Dakota Clutter
Miranda Clutter

7. Anthony Joseph Jaramillo
Donna Cochran, 1st wife
Aaron Jaramillo
Kathy Fisher, 2nd wife
Dylon Jaramillo
Heather Jaramillo

Letter to Pana Abeyta (Sister to Andy Jaramillo)
From: Zenaida (Teacher) First Cousin to
Andy Jaramillo

Tomas Bachicha came from Villanueva New Mexico San Miguel County. Settled in Pinos Wells, NM. Torrance County. Tomas moved to Encino, NM, Torrance County (date unknown). Tomas built or homesteaded his ranch two miles from the little town.

Tomas started out with cattle and sheep, he then homesteaded some more land called El Pinavete (the pine) this was wooded land. Their family made a nice living and raised twelve children.

Tomas still owned land in Villanueva, which my grandmother inherited with lots of fruit trees. We used to go on horse and buggy and a covered wagon to get fruit, which they dried. After your mother married, she learned to canned fruit and taught her younger sisters to can only fruit, vegetables were dried. They also planted lots of corn and had a big garden with lots of good vegetables, but I am sure you remember all that. They dug ditchers to keep cabbage and other vegetables for the winter. There was always milk with good rich cream. The only salad I knew how to eat was lettuce with cream, a little sugar and diluted vinegar, but later mother used to say that we used to have so much meat yet they never learned to cook a roast, they dried it.

When grandfather first got married he was buffalo hunter and used to go to St Louis to trade for goods, they went in caravans. He was 23 or 24 years old when he married grandma, he was ten years older then her.

Tomas (grandfather) was a brilliant man he liked to give speeches for gatherings in Encino, and was Justice of the Peace almost until his death. Grandfather was quite a politician in the Republican Party. After his death we all became democratic. Grandfather ruled with an iron hand. And everybody obeyed. Whole family would knell down to pray the rosary (married and unmarried). During lent we used to pray the station of the cross outside around the homes.

Grandma Baca was a very tiny little woman humble and quite. She had become an orphan at the age of four, but had a precious father who raised his little girl and boy to the best of his ability. Later her father married and they had another son Emiterio. Raphael Baca was grandma's full blooded brother who was killed by the Indians at the age of nineteen. Grandma's most colorful assets were just taking care of her family, God loved both of them. Both grandparents were short on relatives. Grandmother just had two brothers, as her father and mothers the same. Grandpa only had one brother, and Teodoro and one sister, Romana. Both died before grandpa, although he was older. This is all the information I can give you Pana. God love you

Zenaida

APPENDIX B
THE OLGUIN FAMILY

OLGUIN FAMILY
Ancestors of Gillie and Vina Olguin

Jose Maria Olguin
b. 1762
d. Oct. 5, 1796, Tome, Valencia, NM
m. 1786, Belen, Valencia, NM
Maria Francisca Otero, wife
b. 1767, NM

Jose Francisco Rafael Olguin
b. 1796/1797, Jarales, Valencia, NM
m. Dec. 27, 1829, Albuquerque, Bernalillo, NM
Maria Ynez Sedillo, wife
b. May 30, 1802, Atrisco, Bernalillo, NM
Jose Matilde Olguin
b. Mar. 1833, Jarales, NM
d. NM
Juana Maria Garcia, wife
b. Apr. 1842, Belen, Valencia, NM
Jesus Maria Olguin
b. Jan. 16, 1871, Belen, NM
d. Mar. 21, 1938, Belen, NM
Cleotis Sanchez, wife
b. 1873
d. Dec. 3, 1940
Jose Dave Olguin
b. Mar.14, 1898, Jaralez, NM
d. Sept. 15, 1985, Albuquerque, NM
m. Oct. 6, 1919, Belen, NM
Manuelita Gabaldon, wife
b. Dec. 24, 1900, Belen, NM
d. Dec. 18, 1974, Alburqueque
Gillie Albert Olguin
b. April 4, 1929, Belen. NM

Tircio Chaves
Adelaida Corrales, wife of Tircio Chavez
Juan B. Chavez
b. 1880

Pablo Sanchez
Stepanita Aragon
Juan Sanchez
Beatrice Otero, wife of Juan Sanchez
Aruelia M. Sanchez, wife of Juan B. Chavez
b. 1887
Vina Susana Chavez, wife of Gillie Olguin
b. Aug. 11, 1928, Pinos Wells, NM
d. Oct. 17, 1999, Claremont, CA

Vina Susana Chavez and Gillie Olguin married in Belen, NM on Nov. 12. 1949.

Juan Lopez Villasana was born in Fuente Ovejuna, Estremadura, Spain. His son Juan Lopez Holguin was born the same town in 1560/1562, married Catalina Villanueva, came to New Mexico as early Spanish colonists. Their son Cristobal Holguin was born in Santa Fe, NM, in 1617. Cristobal married Melchora de Carvajal. Their son Salvador Holguin, born in NM in 1637, married a Sante Fe lady Magdalena Fresqui, who died in 1695. Their son's last name got changed to Olguin, Juan Lopez Olguin, born in Santa Fe 1655/1660, married Ana Maria Lujan in 1681, in El Paso, when they were refugees from the Pueblo outbreak. Their son Salvador Olguin, born 1682, married Lucia Nieto from Sante Fe, in Nov. 13, 1705. They had Bartolome Olguin, born in Santa Fe in 1706/1710 and who married Maria Romero. They in turn gave birth to Bartolome Olguin, Charma, Rio Arriba, NM, July 19, 1730, who gave birth to Jose maria Olguin [see above]. **Thirteen generations of Olguins in New Mexico, 300 years in what is the present United States.**

DESCENDANTS OF GILLIE AND VINA OLGUIN

Olguin Family Descendants

Children of Gillie 1929 and Vina Olguin 1928-1999

1. Rose Elaine Olguin Villareal, 1952
2. Gary Anthony Olguin, 1955
3. Lorraine Virginia Olguin Sanchez, 1957
4. Ruben Richard Olguin, 1959
5. Michael Andrew Olguin, 1961
6. Natalie Sue Olguin Muckley, 1962
7. John Albert Olguin, 1965
8. DeAnne Charlene Olguin Williamson, 1970

1. Rose Elaine Olguin Villareal
Rudy Raymond Villareal, husband
    Marissa Suzanne Villareal
    Michael Martin Tafoya, husband
        Calob Michael Villareal (Tafoya)
    Melanie Elaine Villareal
2. Gary Anthony Olguin
Nancy Rae Grubb Olguin, wife
    Shane Anthony Olguin
3. Lorraine Virginia Olguin Sanchez
Ralph Ramon Sanchez, husband
    Janesse Renee Sanchez
    Lauren Rose Sanchez
4. Ruben Richard Olguin
Patty Ann Zamora, 1st wife
    Ian John Olguin
    Celese Olguin
Leslie Thomas, 2nd wife
    Garrett Thomas
    Evan Thomas
    Madison Thomas
5. Michael Andrew Olguin
Theresa Gonzalez, wife
    Sophia Susanna Olguin
6. Natalie Sue Olguin Muckley
David Andrew Muckley, husband
    Jenica Lorraine Olguin

Kristen Michelle Muckley
Scott Christopher Muckley
7. John Albert Olguin
Marisa Joy Snyder, wife
Hannah Elisabeth Olguin
Zackary Michael Olguin
8. DeAnne Charlene Olguin Williamson
Greg Williamson, husband

APPENDIX C
THE YSLAS FAMILY

YSLAS FAMILY ANCESTORS
of
Paula Petra and John Yslas

Jesus Yslas
Pomposa Elizalde, wife
Petra Yslas,
b. 1883, Yahualica, Jalisco, Mexico
d. 1921, Clovis, NM

Valentine Yslas
Paula Villegas, wife
Pedro Yslas
b. 1875, Yahualica, Jalisco, Mexico
d. 1930, Clovis, NM
Petra and Pedro Yslas married in Mexico, came to the United States during 1908-9, worked for Santa Fe, then went back to Mexico. After the revolution, they returned with a passport, September 30, 1920. (By 1920 passports were required.)
John Yslas
b. June 24, 1920, La Espada, Jalisco, Mexico
Juan Rincon
b. June 24, 1884, Guadalajara, Mexico
d. 1964, Seligman, Arizona
Celestina Ramirez, wife
b. April 6, 1895, Parral, Mexico
d. 1969, Seligman, Arizona
Paula Petra Rincon
b. June 29, 1920, Seligman, Arizona

John Yslas and Paula Petra Rincon were married in Seligman, AZ, on December 20, 1941, and came to Barstow in 1943.

Yslas Descendants

John Thomas Yslas and Paula Petra Rincon Yslas
Children of John and Paula Yslas:

1. Katherine "Katy" Marie Yslas-Yent
2. John T. Yslas Jr.
3. Frederick J. Yslas
4. Peter Ray Yslas
5. Paul A. Yslas
6. Elizabeth A. Yslas Orloff
7. Andrew J. Yslas

1. Katherine "Katy" Marie Yslas-Yent
Vernon E. Williams, 1st husband
Vernon "Flip" Williams Jr.
Vincent E. Williams Sr.
Joann Hernandez, wife
Vincent E. Williams Jr.
Kyle M. Williams
Steven K. Williams
Vickie Vickers Williams, 1st wife
Nolan R. Williams
Paula B. Williams
Tracy Gaddie Williams, 2nd wife
Allison N. Gaddie
Kayla M. Williams
Matthew S. Williams
Rufus L. Yent, 2nd husband
Derrick G. Yent
Destin S.Yent
2. John T. Yslas Jr.
Dian M. Hixson Yslas, wife
John T. Yslas, the third
Frank L. Yslas
Jeanette M. Yslas
3. Frederick J. Yslas

Cindi K. Wilson Yslas, 1st wife

Marysa K. Yslas

Matthew J. Yslas

4. Peter Ray Yslas

5. Paul A. Yslas

Wendy Fiscus, 1st wife

Phil T. Webster

Lucas T. Yslas

6. Elizabeth A. Yslas Orloff

Robert Nicholas "Bob" Orloff, husband

Nicholas Boris "Nick" Orloff

7. Andrew J. Yslas

Diane M. Miller, 1st wife

Cory A. Yslas

Stevie A. Yslas

APPENDIX D
THE LUNA FAMILY

LUNA FAMILY
Ancestors of Noberto and Annie Luna

Jesus Maria Luna
Maria Rosario, wife of Jesus Luna
    Jose P. Luna (Paz or De LaPaz)
    b. March 19, 1882, Los Lunas, NM
    d. July 6, 1963, Los Lunas, NM
    m. Dec. 14, 1911
Melquedes Gallegos
Filomena Gallegos, wife of Melquedes
    Margarita Gallegos Luna, wife of Jose P. Luna (Paz or DeLaPaz)
    b. June 14, 1891, Los Lunas, NM
    d. July 27, 1971, Albuquerque, NM
        Noberto Luna
        b. May 5, 1916, Los Lunas, NM
        d. Dec. 19,1999, Barstow, CA
        m. Oct. 17,1938, Los Lunas, NM

    Jose S. Zamora
    Josephita Gabaldon, wife of Jose Zamora
        Anita Zamora, wife of Noberto Luna
        b. Feb. 27, 1918, Tome, NM
        d. March 25, 2001, Barstow

Noberto Luna Married Anita Zamora in Los Lunas, NM on October 17, 1938.

LUNA FAMILY DESCENDANTS

Children of Noberto and Annie Luna

1. Leo A. Luna (b.1939)
2. Benny Luna (b. 1941)
3. Lydia Luna (b.1942-d.1943)
4. Helen Luna Raster (b. 1943)
5. Patsy Luna Biles (b. 1945)
6. Nobie H. Luna (b.1946)
7. Randy F. Luna (b. 1951)

1. Leo A. Luna
GJordene Dennis Luna (1st wife) (1941)
    Neil Gregory Luna (1960)
    Winnie Chau, wife (1960)
    Kristine Luna Bryniarski (1963)
    Bruce Bryniarski, husband (1966)
        Daniel Bryniarski (1994)
        David Bryniarski (1995)
    Stephanie Luna Gallop (1965)
    Kevin Gallop, husband (1969)
        Rachel Gallop (1990)
        Nicholas Gallop (1991)
Janet Shattuck Luna (2nd wife) (1952)

2. Benny Luna
Ann Rodriguez Luna, wife (1941)
    Michael Jerome Luna (1961)
    Anita Saiz Luna, wife (1964)
        Isaac Luna (1987)
        Vincent Luna (1989)
        Nicholas Luna (1993)
    Anthony Luna (1963)
    Steven Luna (1967)
    Lottie Acosta Luna, wife (1966)
        Ivan Luna (2000)
    Regina Luna Badillo (1964)
    Darren Badillo, husband (1964)
        Nicole Luna (1982)

Amanda Badillo (1988)
Lucas Badillo (1993)

4. Helen Luna Raster
Harold Raster, husband (1939)
Hal Jacob Raster (1963)
Renee Raster (1983)
Jesse Raster (1986)
Alex Raster (1991)
Van Raster (1967)
Marisa Lopez Raster, wife (1969)
Mathew Raster (1989)
Tatiana Raster (1994)
Landon Raster (1997)
Shannon Raster (1978)

5. Patsy Luna Biles
Albert Gerardi, (1st husband) (1941)
Mark Gerardi (1962)
Dena Hibbetts Gerardi, wife (1961)
Erik Chavez (1980)
Shannon Barkoci (1988)
Cynthia Gerardi McCullough (1963)
Lonnie McCullough, husband (1962)
Stephanie McCullough (1988)
Caleb McCullough (1992)
Michelle Gerardi Jensen (1964)
David Jensen, husband (1970)
Zachariah Gerardi (1986)
Kuylea Jensen (1994)
Haeleigh Jensen (1993)
Jett Jensen (1997)
Jim Biles (2nd husband) (1944)
Chris Biles (1982)

6. Nobie Luna
Nora Tapia Luna (1st wife) (1947)
Jessica Luna-Oliver (1970)
Jimmy Oliver, husband (1971)
Jesse Hernandez (1989)
James Oliver III (1994)
Noby Gabriel Luna (1974)

Analise Luna (1991)
Justina Luna (1994)
Noberto G. Luna (1999)
Manuel C. Luna (2001)
Bernadette Luna (1978)
Romey Santini (1997)
Izabella Luna (2000)
Amber Luna (2nd wife) (1955)
Monica Luna (1980)

7. Randy Luna
Amber Stovall Luna (1st wife) (1955)
Saleta Luna (1971)
Lydia Rodriguez (1994)
Lilliana Rodriguez (1996)
Hilda Silva Luna (2nd wife) (1952)
Alana Price-Brooks (1980)
Colby Price-Brooks (1998)
Debra Hendrickson Luna (3rd wife) (1951)

APPENDIX E THE ARREDONDO FAMILY

ANCESTORS OF JOSE AND HERMENIA ARREDONDO

Jose Maria Contreras
Maria Dolores Contreras, wife of Jose Contreras
    Rafael Contreras

Ramon Zuniga
Maria Razo Arredondo, wife of Ramon Zuniga
    Paula Razo Arredondo Zuniga, wife of Rafael Contreras
        Jose Contreras Arredondo
        Guadalupe Contreras Arredondo, sister
        Miguel Contreras Arredondo, half brother to Jose and Guadalupe
Teodoro Hernandez
Jesus Hernandez
Merced Guerro
Teodora Archiaga Guerro, wife of Merced
    Bartola Lopez Archiaga, wife of Jesus Hernandez
        Alejandro Hernandez
    Pilar Jimenez
    Guadalupe Jeminez Abrego, wife of Pilar Jimenez
        Micaela Abrego Hernandez, wife of Alejandro Hernandez
            Hermenia Hernandez Arredondo, wife of Jose Contreras Arredondo

Jose Contreras took his mother's name "Arredondo" for his last name in the United States. Jose Arredondo married Hermenia Hernandez in 1910, El Paso, Texas. Guadalupe Contreras Arredondo was the aunt who was always a part of the Jose and

Hermenia Arredondo family. Miguel also used the name Arredondo and later lived with Jose and Hermenia until he died.

DESCENDENTS OF JOSE AND HERMENIA ARREDONDO

1. Elvira Arredondo
2. Jesse Arredondo (1914 to 1932)
3. Paula Arredondo
4. Hermenia "Minnie" Arredondo
5. Rafaela Arredondo
6. Annie Arredondo
7. Ramon Arredondo
8. Ruben Arredondo

1. Elvira Arredondo Salgado
Antonio Salgado, husband
    Paula Elinor Conner
    Chris Conner, husband
        Christopher Conner
        Anthony Conner
        Paula Conner
3. Paula Arredondo Grandei
Elwood Grandei, husband
    Linda Grandei Parson
    George Parson, husband
        Jimmy Weisinger
            Bailey Weisinger
4. Minnie Arredondo Carr
Bud Carr, husband
    Buddy Joe Carr
    Sharon Carr, wife
        Tara Carr
            Josephine Martinez
            Andrew Martinez
            Christopher Martinez
        Matt Carr
        Joey Carr
5. Rafaela Arredondo Schultz
Robert Schultz, husband
    Barbara Schultz Lewis

John Lewis, husband
Robert Schultz Jr.
6. Annie Arredondo Vallis
George Vallis, husband
Ralph Vallis
Connie Vallis, wife
Damian Vellis
George Vallis III
Francis Vallis, wife
Alyice Vallis
Ashley Vallis
7. Ramon Arredondo
Esther Arredondo, wife
Alejandro Arredondo
Oscar Arredondo
Paula Lorena Arredondo Langford
Ernie Langford, husband
Brandon Langford
Ryan Langford
Ramon Arredondo Jr.
8. Ruben "Benny" Arredondo
Jacqueline Arredondo, 1st wife
Annette Arredondo Jefferson
Kirk Jefferson, husband
Hunter Jefferson
Jilly Jefferson
Claudette Arredondo Dale
Phil Dale, husband
Logan Dale
Dalton Dale
Jackie Rafaela Arredondo
Taylor Hailey
Liz Arredondo, 2nd wife
Jessica Arredondo
Kimberly Arredondo Lilly

GLOSSARY

Bat out of hell—going very fast
Bizcochitos—cookies, Mexican cookies, originating from New Mexico
CC Camp—Noberto Luna's name for the CCC, i.e., Civilian Conservation Corps, started by President Franklin Roosevelt in the depression of the 1930s.
Coolers—swamp evaporative coolers used in the High Desert
Cuartitos—little shed
Flip off—giving someone the "bird" or the finger
Garaje—garages, the empty space in the mouth when a child loses his teeth
Gandydancers—Anglo name for railroad track workers
Gavacho—slang expression name-calling for a Latino who is "too white" or a gringo or "You think you're too good for us."
Guadalupe—the Virgin of Guadalupe, the first of various miraculous apparitions on Mexican soil. In 1531 Juan Diego, a poor convert, heard heavenly music and a sweet voice calling his name while crossing Tepeyac (hill in Mexico City). The Virgin then appeared, telling him there should be a church built on Tepeyac. A basilica is there now.
Guadalupana—the adjective of name Guadalupe
Las Mananitas—see *Mananitas*
Mananitas—Las *Mananitas*, a early morning song honoring a person on his birthday, usually a surprise. There are many versions to the song, some more religious than others. "Everyone should have this sung to him at least once in his life," said Lorenza Osborn, beloved Spanish teacher retired from Barstow High School. "I had a Las Mananitas for my 21st birthday."
Mariachi—Mexican band
Mi familia—my family
Mi hija or *mija*—my daughter, term of endearment
Mi hijo or *mijo*—my son, a term of endearment
Pozole—corn hominy one dish meal from New Mexico, boiled

with pork, spices and peppers.
Pasole—hominy, see pozole
The projects—San Bernardino Housing (low cost housing, built during the housing shortage in World War II), also Victory Homes
Pelts. Coats of lamb
Rebozos—beautiful bright colored shawls
Ristra—string of red dried chilies
School of Mines—College in Socorro, New Mexico
Route 66—old two-lane highway from Chicago to Santa Monica, named 1926, main road from New Mexico to Barstow
Sonny boy. Andy and Pita Jaramillo's first-born son, Andrew
Tapatias—something from the State of Jalisco, especially from Guadalajara
Trackeros—trackers, track workers, presently called trackers
Trackwalkers—men who checked the tracks to see what repairs are needed
Vieja—old lady or wife, usually meant as an endearment
Viejo—old man or husband, usually meant as an endearment
Victory Homes—County homes in Barstow that housed residents during and after World War II, sometimes called the "Projects"

INDEX

A

Abrego, Guadalupe Jeminez, 195
Abrego, Michaela, 140
Acosta, Lottie, 192
Albuquerque, NM, 6, 19, 121, 179, 180, 184, 191
Amboy, CA, 143, 146, 148, 175
Aragon, Stepanita, 185
Arguellas, Mary Ruth, 28
Arguelles, Alfredo, 23
Arrendondo family, see pgs. 130 -175 and Appendix E
Atrisco, NM, 184
Ayala, Senator Ruben, 167

B

Baca, Congressman Joe, 167
Baca, Emilana, 179
Bachicha, Maria Efrin, 179
Bachicha, Tomas, 179
Badillo, Darren family, 192
Baker, CA, 141
Barstow, CA, 10, 22, 32, 33, 36, 56, 94, 114, 121, 136, 140, 145, 154, 160, 165, 169, 175, 179, 191
Barstow College, 174
Barstow High School, 11, 62, 151, 153, 158, 159, 166
Barstow High School High School, 160
Barstow Junior High, 46
Barstow Police Department, 155
Barstow Union High School, 149
Belen High School, NM, 36
Belen, NM, 10, 32, 36, 47, 184
Bernalillo, NM, 9
Biles, Jim family, 193
Blevins, Miss, 166
Bolo, CA, 143
Bryniarski, Bruce family, 192
Bryniarski, Kristine Luna, 115

C

Cadiz, CA, 143, 145
Cameron School
Barstow, CA, 11
Camp Pond, San Bernardino, CA, 153
Capen, Phil, 166
Carr, Bud family, 196
Carr, William "Bud", 160
Casa del Desierto, 152
Cedillo, Chevo, 158
Chau, Winnie, 192
Chaves, Tircio, 185
Chavez, Dionesa, 179
Chavez, Juan B., 185
Chavez, Juan B. and Aurelia, 37
Chavez, Rufina, 179
Chavez, Vina Susana, 185
Civilian Conservation Corps, 89, 106
Claremont, CA, 185
Clentzos, Pete, 162
Clovis, NM, 69, 188
Conner, Chris family, 196
Connor, Chris, 153
Contreras, 163

Contreras, Jose Maria, 195
Contreras, Maria Delores, 133
Corrales, Adelaida, 185
Corvath, Jeff, 163

D

Daggett Annex, 11, 20
Dale, Phil family, 197
Dale, Philip, 168
De Colores, 121
Del Taco, 40
DeRugoma, Jose Luis, 180
Dolores, Maria, 195
Dr. Graybill
Barstow physician, 11
Duran, NM, 37
Dyson, Susan, 180

E

El Paso, TX, 133, 136, 138, 147, 195
El Santuario Church, NM, 51
Elizalde, Pomposa, 188
Elrod, Major, 33
Encino, NM, 3, 179
Enriquez family, 45
Estrada family, 45
Eucharistic Ministers, 75

F

Father Charles Young, 44
Father Fitzgerald, 11
Fiscus, Wendy, 190
Fisker, Kathy, 16
Fort Bliss, TX, 9, 20
Fort Sam Houston, TX, 10

G

Gabaldon, Josephita, 87, 191
Gabaldon, Manual, 87
Gabaldon, Manuelita, 184
Gallegos, Filomena, 191
Gallegos, Margarita, 86
Gallegos, Melquedes, 191
Gallop, Kevin family, 192
Gallop, Stephanie Luna, 103, 108
gandydancers, 142
garages, 116
Garcia, Juana Maria, 184
Garcia, Nickey, 71
Gerardi, Albert family, 193
Gerardi, Cynthia Marie, 106
Gerardi, Dena, 129
Gerardi, Mark, 101
Goffs, CA, 143
Gonzales, Rufina Muerta, 179
Gonzalez, Jose Antonio, 179
Gonzalez, Theresa, 186
Grandei, Elwood family, 196
Grundei, Elwood, 154
Guadalajara Mexico, 69
Guadalajara, Mexico, 188
Guanajuato, Mexico, 133
Guerro, Merced, 195
Guerro, Teodoro, 195

H

Harper Lake, 170
Harvey, Fred, 66, 84
Hatch, NM, 7
Hector, CA, 143
Henderson Elementary School, Barstow, CA, 57
Hendrickson, Debra, 194
Hernandez, Alejandro, 136
Hernandez, Alejandro family, 195
Hernandez, Herminia, 133, 136
Hernandez, Jesus, 195

Hernandez, Joann, 189
Hernandez, Maria Mora, 175
Hernandez, Queta, 175
Hernandez, Teodoro, 195
Hernandez, Vincente, 140
Hillsboro, NM, 179
Hinkley, CA, 140, 147, 148, 153, 158, 159, 160, 166, 169, 175
Hinkley School, 151, 166
Holguin family, (Olguin), 185
Holguin, Juan Lopez
original name of Olguin, 36
Holmes Supply Co., 145
Horkkanen, Joel family, 181
Howard, Chief, 156
Howard, Floyd, 154
Hughes, Mike family, 181
Hunting Trip, 105
Hurtado, Angelina, 13
Hurtado, Angie, 180

I

Irapuato, Mexico, 131, 133, 165

J

Jarales, NM, 184
Jaramillo family, see Jaramillo Story, pgs. 1 - 30 and Appendix A
Jefferson, Kirk, 168
Jefferson, Kirk family, 197
Jensen, David, 124
Jensen, David family, 193
Jensen, Michelle Gerardi, 120
Jimenez, Pilar, 195

K

Katz Restaurant, 155
Kennedy High School
Barstow, CA, 11, 23
Kozar, Andy, 167

L

La Calera, 134
La Espada, Mexico, 188
La Soledad Rancho, 172
Langford, Ernie family, 197
Larson, Buddy, 166
Larson family, 169
Las Cruces, NM, 179
Limebaugh family, 45
Los Amigos Princess, 40
Los Angeles, CA, 145
Los Lunas, NM, 86, 90, 95, 191
Ludlow, CA, 141, 146, 148
Ludlow Station, CA, 136
Luna family, see Luna Story, pgs. 85 - 129 and Appendix D

M

Mañanitas Tapatías, 80
Marine Corps Logistics Base, 30, 33
Marine Corps Supply Depot, 33
Mauldin, Lupe, 26
Mauldin, Mark, 16, 23
McCullough, Cynthia, 117
McCullough, Lonnie family, 193
McCullough, Stephanie, 109
McIntyre, Captain, 155
Mexican Independence Day, 148
Mexican Revolution, 133
Miller, Diane M., 190
Milligan, Anna Maria, 179
Milligan, William Riley, 19,

179
Minneola, CA, 162
Mojave River Valley Museum, 149, 173
Montoya, Sgt. Fernando, 157
Mount Saint Joseph's Catholic School, 48, 76
Mount Saint Joseph's School, 46, 70
Mountaineer, NM, 38
Mrs. Starkey
teacher at Amboy, 147
Muckley, David Andrew family, 186

N

Needles, CA, 145, 148, 167
Newberry Springs, CA, 96, 141, 143
Norman, Danny, 12

O

Odell, Grace Loomis, 162
Olguin Family
see The Olguin Story
pgs. 31 - 59, and Appendix D
Oliver, Jimmy family, 193
Olsen, Wilber, 166
Orloff, Robert Nicholas family, 190
Otero, Beatrice, 185
Otero, Maria Francisca, 184

P

Palmer, Butch, 158
Pancho Villa, 135
Parral, Mexico, 188
Parson, George R., family, 158, 196
Parson, Linda Grundei, 175
Pierce, Chief Max, 155
Pinos Wells, NM, 185
Pleasant, Mayor Ada, 157
Powell, Judge Ava, 155, 157
Purdy, Paul, 19
Purdy, Thomas David family, 181
Purdy, Tom, 23
Purdy, Tommy, 14

R

Ralph's Market, Los Angeles, 145
Ramirez, Celestina, 188
see Yslas family, 64
Raster, Hal Jacob, 102
Raster, Harold family, 193
Raster, Helen, 105, 106, 129
Raster, Matthew, 109
Raster, Van, 107
Renteria, Joe, Sr., 158
Reserve, NM, 179
Rincon, Juan, 188
see Yslas family, 64
Rincon, Paula Petra, 188
see Yslas Family, 64
Rodriguez, Ann, 192
Rosario, Maria, 191
Roswell, NM, 35
Route 66, 95, 96, 98, 145
Roy's Café, Amboy, CA, 144
Rugoma, Nick, 180

S

Sabinal, NM, 179
Saenz, Chelsea, 181
Saiz, Anita Lorraine, 114
Salgado, Antonio, 152, 196
Salgado, Paula Eleanor, 153
San Bernardino, CA, 145, 160

San Bernardino Valley College, 13
San Felipe Catholic Church Albuquerque, NM, 6
San Francisco, CA, 40
San Juan, NM, 179
San Pedro, NM, 179
Sanchez, Aruelia M., 185
Sanchez, Cleotis, 184
Sanchez, Pablo, 185
Sanchez, Ralph Ramon family, 186
Santa Fe Depot, Hinkley, CA, 149
Santa Fe Employees, 141
Santa Fe Railroad, 70, 138, 141, 142, 145, 151, 152, 154, 160, 166, 168, 188
Santa Fe Road Master, 136
Santa Rita, NM, 133
Schultz, Rafaela, 175
Schultz, Robert "Bob", 161
Schultz, Robert family, 196
Section Gang Workers, 142
Sedillo, Maria Ynez, 184
Seligman, AZ, 61, 188
Seligman Grade School, 83
Shattuck, Janet, 192
Shepard, Congressman Harry, 168
Shepard, Congressman Harry R., 170
Shoshone, CA, 141
Silva, Guadalupe "Pita", 180
Silva, Hilda, 194
Silva, Tomas, 179
Simko, Eddie, 15
Simko, Geraldine, 25
Sinner, David family, 181
Sister Dulce Maria, 41
Snyder, Marisa Joy, 187
Socorro, NM, 3, 180
Soutar Dodge, 46
Spain, 36
St. Mary's Catholic School, 36
State of Guanajuanto, Mexico, 131, 152
State of Michuacan, Mexico, 162
State of San Luis Potosi, Mexico, 131, 139, 140
Stovall, Amber, 194

T

Tacubaia, Mexico, 137
Tapia, Nora, 193
Thomas, Leslie, 186
Tome, NM, 87, 184
Tonapah and Tidewater Railroad, 141
Tonopah and Tidewater Railroad, 141
Townsend, Brian, family, 180

U

USO, 157, 159, 161

V

Vallis, Chonita, 162
Vallis, George family, 197
Vallis, George, Jr., 162
Venner, William, 22
Victory Homes, Barstow, CA, 63, 91
Villafana, David, 175
Villanueva, NM, 179
Villareal, Calob, 56
Villareal, Rudy family, 186
Villegas, Paula, 188
see Yslas family, 69
Visalia, CA, 144

W

Walker Air Force Base, NM, 35
Watson, Major, 154
Webster, Phil T., 190
Wilkes Hospital
  Barstow, CA, 11, 25
Williams, David, 180
Williams, M. D., 23
Williams, M. D. family, 180
Williams, Steven, 77
Williams, Vernon E. family, 189
Williams, Vernon Eugene, Jr., 62
Williams, Vincent
  see Yslas family, 76
Williamson, Greg family, 187

Y

Yahualica, Jalisco, Mexico, 69
Yahualica, Mexico, 188
Yent, Katy
  see Yslas Family, 61
Yslas Family
  see Yslas Story pgs. 60 -84 and Appendix C
Yslas, Liz, 46
Yslas-Yent family, 189

Z

Zamora, Anita Feliz, 87
Zamora, Jose P., 87
Zamora, Jose S., 191
Zamora, Patty Ann, 186
Zeluff, Mr. and Mrs., 84
Zuniga, Ramon, 195